Humanitarian Engineering for Global Oncology

Online at: https://doi.org/10.1088/978-0-7503-3751-9

IOP Series in Global Health and Radiation Oncology

Series editor
Wilfred Ngwa
Dana-Farber/Harvard Cancer Center, University of Massachusetts Lowell, Johns Hopkins University, USA

Editor biography
Wilfred Ngwa is the Director of the Global Health Catalyst, a cross-institutional collaboration initiative launched at Harvard to catalyse high impact collaborations in global health. He currently serves as Adjunct Professor at the University of Massachusetts, as Associate Professor of Radiation Oncology at Johns Hopkins University and as ICTU Distinguished Professor of Public Health. He is a chair of the Lancet Oncology Commission for Sub-Saharan Africa, has published three books on global health and serves on the editorial board of a number of journals, including ASCO's *Journal of Global Oncology, Ecancermedicalsciences*, and *Frontiers in Oncology*. He has won many awards from Harvard, the USA National Institutes of Health, and international professional organisations for his innovations and leading work in global health to address disparities in the USA and globally.

Aims and scope
This series includes books in the emerging area of global radiation oncology and its applications in global health. Building on the published book by the series editor entitled *Emerging Models for Global Health in Radiation Oncology*, it will further detail the work being done globally to promote cancer research and awareness, particularly in lower-income countries.

A full list of titles published in this series can be found here: https://iopscience.iop.org/bookListInfo/iop-series-in-cancer-research-for-global-radiation-oncology#series.

Humanitarian Engineering for Global Oncology

Edited by
Eric Ford
Department of Radiation Oncology, University of Washington, Fred Hutch Cancer Center, Seattle, WA, USA

IOP Publishing, Bristol, UK

ISBN 978-0-7503-3751-9 (ebook)
ISBN 978-0-7503-3749-6 (print)
ISBN 978-0-7503-3752-6 (myPrint)
ISBN 978-0-7503-3750-2 (mobi)

DOI 10.1088/978-0-7503-3751-9

Version: 20240601

IOP ebooks

British Library Cataloguing-in-Publication Data: A catalogue record for this book is available from the British Library.

Published by IOP Publishing, wholly owned by The Institute of Physics, London

IOP Publishing, No.2 The Distillery, Glassfields, Avon Street, Bristol, BS2 0GR, UK

US Office: IOP Publishing, Inc., 190 North Independence Mall West, Suite 601, Philadelphia, PA 19106, USA

To our patients and their families

Contents

Editor biography

Eric Ford

Eric Ford, PhD FAAPM FAIMBE FASTRO is a Professor at the University of Washington/Fred Hutch Cancer Center, Seattle and Vice-chair and Director of Medical Physics in the Department of Radiation Oncology. He is an international leader in cancer care with a focus on radiation oncology physics, experimental radiation biology, and global oncology. He is an active researcher with over 160 published papers and three books and has been principal investigator of multiple federal grants. He is a frequent invited lecturer nationally and internationally and is a valued mentor to students and trainees recognized with six teaching awards.

List of contributors

Beatrice Wiafe Addai
Peace and Love Hospitals and Breastcare International, Ghana

Lawrencia Dsane Bawuah
Erasmus MC Cancer Institute, University Medical Center Rotterdam, Rotterdam, The Netherlands
and
Consultant Peace and Love Hospitals, Ghana

Ross Berbeco
Department of Radiation Oncology, Brigham and Women's Hospital, Dana-Farber Cancer Institute and Harvard Medical School Boston, MA, USA

Derek Brown
Department of Radiation Medicine & Applied Sciences, UC San Diego Health, La Jolla, CA, USA

David Butler
National Academy of Engineering, The National Academies of Sciences, Engineering, and Medicine, Washington DC, USA

Daniel T Chiu
Department of Chemistry and Bioengineering, University of Washington, Seattle, WA, USA

Gabriel Conzuelo
Basic Health International, Pittsburgh, PA, USA

Laurence E Court
Department of Radiation Oncology, University of Texas MD Anderson Cancer Center, Houston, TX, USA

Miriam Cremer
Basic Health International, Pittsburgh, PA, USA

Claire Dempsey
Department of Radiation Oncology, Calvary Mater Newcastle Hospital, School of Health Sciences, University of Newcastle, Waratah NSW, Australia

Eric Ford
Department of Radiation Oncology, University of Washington, Fred Hutch Cancer Center Seattle, WA, USA

Skylar S Gay
Department of Radiation Oncology, University of Texas MD Anderson Cancer Center, Houston, TX, USA

Lensa Keno
Department of Health Administration and Human Resources, The University of Scranton, Scranton, PA, USA
and
Johns Hopkins University, Baltimore, MD

Guru Madhavan
National Academy of Engineering, The National Academies of Sciences, Engineering, and Medicine, Washington, DC, USA

Rachel Masch
Basic Health International, Pittsburgh, PA, USA

Michele Moreau
Department of Radiation Oncology, Brigham and Women's Hospital, Dana Farber Cancer Institute, Boston and Harvard Medical School, Boston, MA, USA

Marios Myronakis
Department of Radiation Oncology, Brigham and Women's Hospital, Dana-Farber Cancer Institute and Harvard Medical School Boston, MA, USA

Saurabh S Nair
Department of Radiation Oncology, University of Texas MD Anderson Cancer Center, Houston, TX, USA

Wilfred Ngwa
Department of Radiation Oncology, Johns Hopkins University, Baltimore, MD, USA

Joshua S Niedzielski
Department of Radiation Oncology, University of Texas MD Anderson Cancer Center, Houston, TX, USA

Jerald P Radich
Clinical Research Division, Fred Hutch Cancer Center, Seattle, WA, USA

Adam Shulman
UC Health, Colorado Springs, CO, USA

Montserrat Soler
Basic Health International, Pittsburgh, PA, USA

Sayeda Yasmin-Karim
Department of Radiation Oncology, Brigham and Women's Hospital, Dana Farber Cancer Institute, and Harvard Medical School, Boston, MA, USA

Afua A Yorke
Department of Radiation Oncology, University of Washington, Fred Hutch Cancer Center Seattle, WA, USA

Jiangbo Yu
Department of Chemistry and Bioengineering, University of Washington, Seattle, WA, USA

Jicheng Zhang
Department of Chemistry and Bioengineering, University of Washington, Seattle, WA, USA

IOP Publishing

Humanitarian Engineering for Global Oncology

Eric Ford

Chapter 1

Introduction to humanitarian engineering: varieties of technical experience

David Butler and Guru Madhavan

Over a century ago, Morris Llewellyn Cooke—best known for his later work on rural electrification—sought to lay the groundwork for a code of ethics for the engineering profession in the United States of America. His bottom line in this effort was unequivocal: 'The ultimate goal here is the flatfooted declaration that good engineering must be in the public interest and, contrariwise, that any engineering which is anti-social must be bad engineering' [1].

The statement derives its power from its simplicity. Good engineering is commonly thought of as economical, elegant, and effective. While it is all those things, it is not only those things. Cooke, who was practicing in a world still recovering from the effects of the devastating flu pandemic, recognized that the profession also owed society an obligation to ensure that the wonders it brought forth benefited everyone.

But how might that obligation be realized? One way is to adopt an approach that has gained attention in recent years: engineering for a *just future*, a concept with foundations in humanitarian engineering that will feel familiar to anyone involved in global oncology. There are various articulations of it, but four interconnected principles dominate:

Cultural awareness that privileges a design perspective factoring the diverse needs, preferences, and backgrounds of the end users and incorporating appropriate features or adaptations to ensure inclusivity. Engineers evaluate the potential impact of their designs and projects on local communities, marginalized groups, and indigenous populations.

Ethical integrity, through which engineers are truthful and transparent in their professional dealings, exhibit fairness and impartiality in decision-making processes, and prioritize the health, safety, and welfare of the public and the environment in their work.

doi:10.1088/978-0-7503-3751-9ch1

Social accountability that compels engineers to actively collaborate with those from diverse cultural, professional, and experiential backgrounds to gain the broad perspective needed to address societal challenges effectively. Research shows that fostering diversity within engineering teams and creating inclusive work environments leads to more equitable solutions.

Environmental Responsibility, reflected through a commitment to fulfill a project's functional needs in a manner that minimizes the cost to the customers and communities for building, manufacturing, operating, distributing, and ultimately managing a system's end of useful life. Designing for sustainability, resilience, minimizing the adverse effects on ecosystems and communities, and responsible resource management are all parts of this effort.

Together, these 'CESER' principles provide the foundation for a holistic approach to engineering complex systems that integrates the consideration of *people*, *systems*, and *culture* and seeks to achieve just outcomes. Such considerations also lie at the core of global oncology. Global oncology acknowledges the existence of cultural diversity and promotes approaches to cancer care that are respectful of cultural norms, values, and preferences, fostering trust and improving patient outcomes. It actualizes ethical principles in cancer prevention, treatment, and research, providing access to care without regard to socioeconomic status, geographic location, or cultural background and ensuring that studies examine the full range of people affected by the disease. Global oncology recognizes the social justice imperative to address health disparities, achieve health equity, and empower local communities and healthcare providers. Additionally, it encourages the adoption of environmentally responsible health-based sustainable practices as part of a comprehensive approach to cancer prevention.

Cancer care in the global context is in a period of rapid growth. The number of new cancer cases worldwide grew from an estimated 12.7 million in 2008 to 19.3 million in 2020 [2]; by 2040 this number will likely rise by nearly 50% over the 2020 estimate, with the largest increases in countries with transitioning economies [3]. While there are many factors driving this growth, much of it stems from rapid societal and economic transformations in many low- and middle-income countries as well as major recent advances in managing communicable diseases. As such, the burden of disease is now tipping towards non-communicable diseases such as cancer. In the coming decade cancer will likely become the leading cause of death not only in high-income countries but across the globe.

Navigating this change and growth will require a multi-pronged approach in which engineering will play a key role. Why is engineering important for global oncology? One prominent reason is the technological advances that engineers bring about. Devices and associated technology are a central component in the delivery of quality cancer care, ranging from prevention, to diagnosis, to treatment of cancer. Whether it be a telemedicine app that allows people in remote locations to access quality advice and care, a portable MRI scanner that facilitates diagnoses in communities too small to support a dedicated unit, or an inexpensive cooler that keeps medications at their proper temperature for extended periods in places where reliable power is not a given, engineers are at the forefront of developing tools that help to make global oncology possible.

This is where CESER principles come in. Engineering projects are constrained not just by the requirements and budget assigned to them, but by the people who are tasked with implementing them. If design teams lack people with knowledge of the everyday challenges of all of the potential users of their product and the strength of character to apply that knowledge when faced with resistance from those for whom the monetary bottom line is the only criterion, they will not have the wherewithal to make the choices that allow technology to work in every environment where it may be used and serve the needs of the people in those diverse environments.

While centering on engineering *for* people, systems, and culture is vital, it is insufficient. Just as importantly, attention must be given to the people, systems, and culture *of* engineering so that the profession itself is receptive to the more diverse perspectives, philosophies, and professional practices essential for tackling global health problems. The consequences of not giving these factors due consideration are many. One classic example of an engineering failure in this respect is the New York State Thruway system. To be clear, the system is a marvel of engineering, linking the state with a modern, time-saving roadway that greatly benefited commerce. It 'succeeded', however, at the expense of many poor and minority communities in its path, which were ripped asunder without so much as an entrance and exit ramp in compensation. The Thruway created physical barriers that divided neighborhoods and disrupted social cohesion and economic opportunities while increasing air and noise pollution. Its benefits flowed primarily to those who were already advantaged, while further disadvantaging those least able to bear the burden.

How can this goal be realized? Education is naturally part of the answer, but the CESER principles mean nothing if they are not translated from paper or the classroom to the real world. Although the concept of teaching character has been around at least since the time of Plato, it has only recently gained currency as a component of engineering education. At its root, it entails arming students with the basics of professional responsibility—recognizing conflicts of interest, complying with regulations and standards, identifying illegal and unethical behavior, and the like. However, comprehensive character education goes beyond this, 'cultivating stable dispositions to act, think, and feel in ways that enable us to do the right thing, for the right reasons, in the right ways', no matter the particulars of the given situation [4]. Making such principles a fundamental part of education and practice is essential to engineering for a just future. As more and more character-trained engineers sensitive to the greater cultural, social, ethical, and environmental implications of their work become practitioners and take leadership roles, they will be able to remake the baseline assumptions of the profession and implement the changes needed to move it forward.

The following chapters of this book explore and probe these problems further. They discuss the complexities present in meeting the changing needs of global oncology care. The authors present examples of well-established technologies that are now undergoing further refinements, and they also present examples of emerging technologies such as novel nanomaterial fluorescent probe assays for cancer biomarkers and other applications. But just as importantly, the chapters in this text focus not only on the technology itself but the design and deployment of the technology. They engage with economics, sociology, and philosophy of care, aspects core to engineering design. The chapters should ultimately allow readers to reflect on the vital role engineering plays between the porous borders of capitalism and community and care and commercialization. After all, as creators and consumers of engineered systems, it remains to be seen whether our solutions are more in the vein of 'good engineering...in the public interest' (as Cooke put it) or 'anti-social...bad engineering'.

References

[1] Cooke M L 1922 Ethics and the eingineering profession *Ann. Am. Acad. Pol. Soc. Sci.* **101** 68–72

[2] Bray F, Jemal A, Grey N, Ferlay J and Forman D 2012 Global cancer transitions according to the Human Development Index (2008–2030): a population-based study *Lancet Oncol.* **13** 790–801

[3] Sung H, Ferlay J, Siegel R L, Laversanne M, Soerjomataram I and Jemal A *et al* 2021 Global Cancer Statistics 2020: GLOBOCAN estimates of incidence and mortality worldwide for 36 cancers in 185 countries *CA Cancer J. Clin.* **71** 209–49

[4] Pierrakos O, Prentice M, Silverglate C, Lamb M, Demaske A and Smout R 2019 Reimagining engineering ethics: from ethics education to character education *2019 IEEE Frontiers in Education Conf. (FIE) (Covington, KY, 16–19 Oct)*

Chapter 2

Cost-effective radiation treatment delivery systems for low- and middle-income countries

Afua A Yorke and Eric Ford

With cancer becoming an increasing burden of disease in low- and middle-income countries (LMICs) it is increasingly important to pursue affordable, sustainable solutions for cancer care. Radiotherapy is one component of this disease management spectrum and has been shown to be affordable and feasible and can be consistently deployed in LMICs. However, radiotherapy centers in LMICs are home to only 5% of the world's radiotherapy resources. The result is delayed access to care (ranging from 4 months to 17 months), advanced disease at presentation, and increased mortality. While there are many dimensions to the issue of radiation therapy (RT) access, one key component centers around the technology in use and the associated cost and complexity. This chapter focuses on the cost effectiveness of RT delivery technologies. This includes linear accelerators (linacs) and high dose rate (HDR) brachytherapy because these are the main modes of delivering treatments in LMICs. In addition, we discuss other emerging technologies with the promise of providing cost-effective RT care.

2.1 Introduction

In the coming years the number of worldwide cancer cases is projected to increase, becoming an urgent global problem. By 2030 reports estimate there will be 27 million new patients and more than two-thirds of cancer-related deaths will occur in LMICs [1], imposing the greatest challenge on cancer management in these countries where the greatest burden of diseases lies with non-communicable diseases such as cancer. The rising burden of cancer in LMICs adds to the stress of existing weak healthcare and economic systems.

In addition, cancer care treatments when available present a financial burden to patients. The issue of financial toxicity has been well explored in cancer practices in higher-income countries (HICs) [2]. Unlike most patients, cancer patients may

doi:10.1088/978-0-7503-3751-9ch2

receive multiple treatments including surgery, RT, and systemic treatment. A component of financial hardship that influences the vulnerability to financial distress is the household income of the patient, spouse, or family. For cancer patients in LMICs this may be the most important factor to consider.

Given the above considerations, it is more important than ever to pursue affordable cancer care in LMIC settings especially given the rapid development of new technologies that are contributing to driving up the cost of care. RT is one of the most cost-effective forms of therapy for cancer treatment since it is noninvasive, and due to the long lifespan of the therapy machines and the large volume of patients that can be treated over the lifetime of a therapy machine. Radiotherapy has been shown to be affordable and feasible and can be consistently deployed in LMICs. The investment in radiotherapy can yield substantial gains in all income groups [3]. The availability of an RT facility and the adequate access to care are very important in the management of cancer and thus have a major impact on the quality of healthcare provided. RT machines are equipment needed to provide radiation treatment. In LMICs RT delivery devices include linacs or cobalt-60 units, as well as HDR and low dose rate brachytherapy units. RT capabilities have grown rapidly over the past decades in many regions due to increased governmental awareness of the importance of RT in the management of cancer (see for example figure 2.1 which shows the growth of linac installations in Africa) [4, 5].

However, there are many regions lacking in RT resources. Although it has been estimated that 80% of the world's cancer burden is in LMICs, RT centers in these regions are home to only 5% of the world's radiotherapy resources [6]. Furthermore

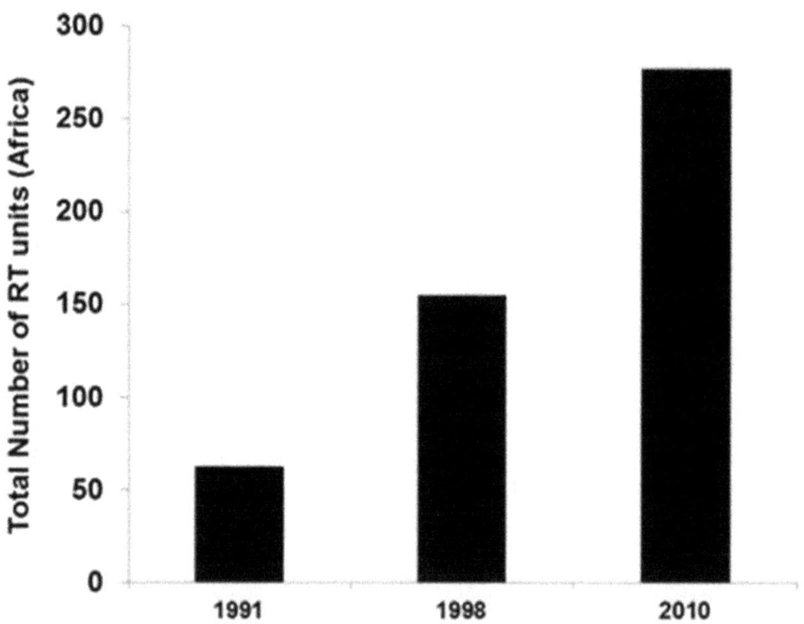

Figure 2.1. Growth of radiotherapy in Africa. Using data from [5].

there are unique operational challenges to delivering RT in LMICs because translation of radiotherapy technologies developed in HICs to LMICs often does not work as intended.

The result of limited access to critical resources for cancer management leads to delayed patient care due to delayed diagnosis which may lead to more advance disease at presentation and increased mortality [7]. Studies shows that patient delays in Africa range from 4 months to 17 months [8]. A recent study from one clinic in northern Nigeria showed that the median wait times from radiotherapy consult to first treatment was 130.5 days, which is far longer than recommended for disease control [9].

While there are many dimensions to the issue of RT access, one key component centers around the technology in use and the associated cost and complexity. This chapter will focus specifically on the cost effectiveness of RT delivery technologies. This includes linacs and HDR brachytherapy given these are the main modes of delivering treatments in LMICs. In addition, we discuss other emerging technologies with the promise of providing cost-effective RT care. Treatment planning and other aspects of the care pathway are the subject of other chapters.

2.2 Overview of RT delivery technologies: needs and costs

It has been reported that approximately 50% of cancer patients will undergo radiation treatment during the course of their diagnosis [9]. Radiotherapy is noninvasive and, in the patients that require it, it allows for organ preservation with low risk of morbidity and overall low cost. In recent years progress has been made towards the treatment delivery with the advent of technological advances which have allowed for the addition of imaging modalities, software, and modern linacs.

2.2.1 Technology needs (external beam radiation therapy)

Radiation delivered to the tumor can be administered in two ways: (1) external beam radiation therapy (EBRT) and (2) brachytherapy. EBRT is delivered from the outside of the body by directing high-energy particles, i.e. photons, electrons, protons, or other heavy ions, to the targeted tumor. Brachytherapy on the other hand is delivered from the inside of the body usually using sealed radioactive sources. EBRT can be delivered with a simpler technique using three-dimensional conformal radiation therapy (3DCRT) or more complex techniques such as intensity-modulated radiation therapy (IMRT) or volumetric modulated arc therapy.

The technological advancement in radiotherapy plays an important role in the quality of care patients receive and improved outcomes. IMRT techniques allow for greater organ sparing and limit tissue toxicity which can be costly and toxic to manage. For example, given the proximity of head and neck cancer to the spinal cord, parotid glands, eyes, lacrimal glands, optic nerves, and auditory structures, IMRT can be used to avoid these organs at risk while still being able to treat the target volume with curative doses. This limits the development of acute and late toxicities such as xerostomia which increases the chance of infection, ulcers, and osteoradionecrosis [10–12]. The clinical relevance of the advantage of IMRT over 3DCRT has been investigated and is in widespread use for various disease sites [13]. As a result, IMRT is offered in essentially every center in North America and Europe and yet in many LMICs RT is only available in the simpler 3DCRT form.

In addition to the advancement in IMRT, there have been other techniques that have been developed to enable an increase in precision and the accurate delivery of radiation by using tighter margins. One such technique is image-guided radiation therapy (IGRT). When combined with IMRT, IGRT can contribute to the improvement of outcomes by ensuring an accuracy in patient setup for each radiation fraction. IGRT provides the use of hypo-fractionated RT schedules in some disease sites which reduces the treatment time and increases patient throughput which can be a major advantage in the LMIC setting [14].

It should be noted that a comprehensive implementation of IMRT needs a robust infrastructure. Studies indicated that in low-resource settings, power outages and relatively long machine downtime of linacs can have a profound impact on the ability to deliver RT [15]. For example, a 1-hour power outage per day reduces the efficiency of treatments on a linac with step-and-shoot IMRT below 70%.

2.2.2 Technology needs (brachytherapy)

The second major method of delivering RT is through brachytherapy which, in the LMIC context, is driven mainly by the need to treat cervical cancer. Cervical cancer is a growing problem in LMICs and the fourth most common cancer diagnosed in women worldwide and of the 530 000 cases of cervical cancer diagnosed in 2012, 85% occurred in LMICs [16]. Brachytherapy is a critical component of definitive therapy for many patients with cervical cancer who are not candidates for surgery per the National Comprehensive Cancer Network (NCCN). Brachytherapy may be combined with EBRT, in which case the brachytherapy is often initiated towards the end of treatment when sufficient tumor has regressed and will permit brachytherapy apparatus geometry. For example, cylinder brachytherapy is used to treat the vaginal cuff after surgery for early-stage endometrial cancers. Brachytherapy can be delivered using either intracavitary devices or in combination with interstitial devices for more advanced disease. For advanced cervical disease treatments are complex and require interstitial needles for eradicating residual disease following external beam therapy [17]. For very early-stage diseases brachytherapy alone may be used.

Brachytherapy plays an important role in the treatment of certain types of cancers beyond cervix, including prostate, esophagus, and nasopharynx because a radioactive source can be placed in the tumor providing a high dose of radiation to the target while sparing the surrounding sensitive organs.

There are numerous technology needs for delivering HDR brachytherapy. An afterloader device is required to control the administration and treatment with the radioactive sources. Brachytherapy also requires imaging, both to define the extent of disease (with MRI being the recommended gold standard) and to verify the position of the brachytherapy applicator device inside the patient. All of this requires a consideration of the upfront capital and maintenance costs including replacement of the radioactive source which can be frequent depending on the type of isotope. Other considerations include training, reliable power supply, and technical expertise in the individuals operating the afterloader. Hence this requires personnel training on the commissioning and clinical implementation of brachytherapy.

2.3 Principles of cost-effective solutions

To date, many of the technology solutions in radiation oncology have evolved in an organic way and have not been designed with the LMIC environment in mind. They often rely on reliable power grids and infrastructure, highly trained expert users, and near-immediate availability of expert engineers and parts. As RT continues to become widely available in LMICs and as patient throughput is increased given the limited number of machines available, it is becoming important to consider these costs more explicitly in the design of technology.

There are numerous costs involved in the delivery of quality RT. Much of the time, the initial focus is on the upfront capital cost, but it is also important to consider the cost of maintenance (which is often equal to the upfront capital cost of a technology). In addition to the cost of maintenance there is also the cost that is incurred when a device is unable to treat. In some environments where parts of supply chains are weak and engineering service support is challenging the downtimes can be extended. This incurs both a financial cost and a cost in terms of human health.

In addition, there are costs and considerations related to personnel. In many countries it is expected that highly trained staff will be utilizing this equipment. However, there is an acute shortage of oncology professionals in LMICs. For example, there are approximately 700 medical physicists in Africa which translates to 0.6 per million population versus Europe which has 13 per million and United States of America and Canada which have even higher rates [18]. Given this, design choices are even more important. More robust designs of software and hardware that do not rely as much on highly trained users will improve efficiency and reduce error [19, 20]. It is also certain that in the future automation will play an increasingly important role in technology deployment through use of artificial intelligence and other approaches.

In summary technologies that are designed for use in LMICs should consider the unique challenges and associated costs in the realms of capital, reliability, maintenance, and service as well as the human components which include the need to staff clinics with the appropriately trained people and the ability for those people to interact with hardware and software platforms in a way that is reliable and efficient. Although many of the technology solutions in radiation oncology have evolved in an organic way that has not explicitly considered the LMIC environment, in the subsequent sections we will highlight several recent solutions that provide a counterexample.

2.4 EBRT solutions

2.4.1 Challenges with current EBRT technology and emerging solutions

To date the most widespread technology for delivering external beam RT is the conventional C-arm linac first commercialized by Varian Inc. in 1968 as the Clinac-4 system and remains largely unchanged to date. These are complex systems which require large thermal loads to operate successfully. Modern C-arm linacs also have sub- and ancillary systems such as the multileaf collimators, volumetric imaging, robotic couches, real-time motion, and monitoring with infrared, electromagnetic, and optical systems. As a result, a comprehensive quality assurance is required on C-arm linacs for both the main and subcomponents to ensure that current performance parameters have not deviated from baseline values at the time of acceptance of the machine. Also, the beam models in the treatment planning systems that are used in conjunction with the linacs need to be validated periodically which requires highly trained medical physicists to carry out these quality assurance measurements. This can create a burden on the already limited resources in terms of education and training and the low number of physicists working at these centers. There is also the added cost of machine breakdown which results in machine downtime, treatment delays, and pressure on the clinic staff and can increase the risk of clinical errors.

In recent years healthcare equipment manufacturers for radiation oncology devices have designed systems that are more operationally efficient. One example of this is the Halcyon system (Varian Inc.), a ring-based delivery system which features a 100 cm gantry opening larger than a standard CT bore, a linac capable of producing a 6 MV flattening filter-free (FFF) photon beam with a maximum dose rate of 800 MU min^{-1}. A schematic diagram of the Halcyon system is shown in figure 2.2. The Halcyon is capable of rotations four times faster than C-arm gantries for rapid imaging and treatment.

According to a recent study [21] some advantages of the Halcyon system include easy transport and installation (<2 weeks) and increased patient throughput by about a factor of 4. Several design features of the Halcyon platform allow for shorter treatment times and hence a high volume of patients. These include an FFF beam and a higher-speed multileaf collimation system. Additionally, a simplified user interface reduces the complexity of treatment and hence requires less operator training.

There are also several mechanical design features in the platform that appear to offer advantages in terms of power consumption and reliability. The linac operates only at a low energy (6 MV); this makes it simpler in design and more efficient and eliminates the need for a complex bending magnet assembly. The Halcyon also requires 40% less electricity during beam-on state and 30% less electricity in standby state than conventional technology. For LMICs with power fluctuation issues this energy-efficient equipment can provide an advantage. The system is also mechanically more reliable in some ways. The ring gantry provides for an inherently stable design. The multileaf collimator system employs leaves that are twice as thick as most modern linac (10 mm versus 5 mm) which allows for more reliable drive motors to be used. Although this otherwise results in less ability to shape the beam, a

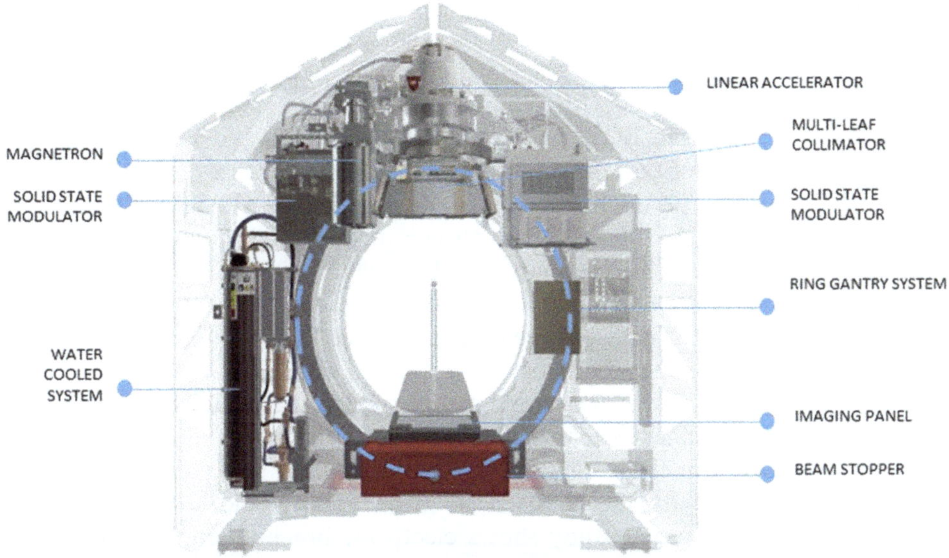

Figure 2.2. A schematic diagram of the Halcyon™ system (Varian Inc.). A low-energy linac, ring design, and large thickness collimator leaves contribute to a more energy-efficient and reliable design. Reproduced from Cozzi L, Fogliata A, Thompson S, *et al* Technol Cancer Res Treat **17**, copyright © 2018 by (Cozzi *et al*). Reprinted by Permission of SAGE Publications. Reproduced from [22]. CC BY 4.0.

double-stack design was developed with one leaf bank offset by 5 mm from the other bank, which provides for a spatial resolution that is similar to other modern linacs.

2.4.2 Specialized delivery systems for EBRT

In addition to the aforementioned, there are other specialized systems such as the CyberKnife and Gamma Knife. These are mainly designed for specialized usage such as in stereotactic radiosurgery (SRS) treatment and so have had a more limited utility in LMICs than the more general platforms discussed above. However, some aspects of the specialized systems that are emerging more recently serve to illustrate cost-effective design solutions. One example is the ICON® system released for the Gamma Knife platform which provided an integrated cone-beam CT imaging system. Prior to that, only frame-based SRS treatments were widely available which limited the types of patients who could be treated and made procedures more complex. Another example is the ZAP-X gyroscopic radiosurgery platform [23, 24]. Although mechanically complex, this system offers the design feature that it is self-shielded for RT which simplifies the installation and use.

2.5 HDR brachytherapy solutions

HDR brachytherapy is an appropriate and necessary treatment modality for many types of cancer including cervical cancer [25]. HDR affords the opportunity for many patients to be treated in one or several days often as outpatients.

A radioisotope with high specific activity is mostly used for HDR brachytherapy. The most used isotope is ^{192}Ir because of its high specific activity (330 MBq mm^{-1}), relatively low gamma energy (average 0.375 MeV), and relatively short half-life. However, the short half-life (74 d) requires the frequent replacement of the HDR source. However, starting in 2005 a new type of HDR source was developed which uses ^{60}Co. This is attractive for radiotherapy centers in LMICs because of the much longer half-life of 5.3 years. Although ^{60}Co has a higher energy (1.25 MeV) than ^{192}Ir dosimetry, studies show [26–28] that it can provide an acceptable dose to organs at risk. Given the long half-life of ^{60}Co the source can be used for up to about 5 years. In this 5 years there would have been about 20 sources exchanges for ^{192}Ir. In other words, ^{60}Co ... is very cost effective for high-volume clinics.

^{60}Co source has well-understood physical properties per TG-43 dosimetric measurements and Monte Carlo calculations and source models are available in commercial products. Studies suggest that the plan quality is similar to that achieved with ^{192}Ir [26–28].

In addition, there are electronic brachytherapy systems like the Xoft® Axxent® system (iCAD inc.) that can be used to treat gynecological cancers such as endometrial cancer. A recent study shows electronic brachytherapy offered equivalent target volume coverage to ^{192}Ir treatment with electronic brachytherapy system and allowed for an increase in organ sparing for bladder and rectum [29]. Unlike conventional HDR systems, the Xoft system does not require any source replacement and needs only minimal shielding which makes it very cost effective for less resourced facilities. The mobility of the system makes it easy to treat patients at multiple locations and can be easily stored when not in use.

There are other challenges for brachytherapy in the LMIC setting associated with imaging needs, especially MRI. The image-guided intensity-modulated external beam radiochemotherapy and MRI-based adaptive BRAchytherapy in locally advanced CErvical cancer (EMBRACE) study initiated by GEC-ESTRO provides benchmark image guidance for brachytherapy. The study recommends MRI as the gold standard. The EMBRACE guidelines recommend brachytherapy treatment planning based on MRI imaging with the applicator in place and additional criteria for MRI sequencing, contouring, applicator reconstruction, and dose optimization [30]. For treatment planning, it is recommended that MRI be performed in the same position as the planned treatment with either MRI obtained with 1.5–3 T scanners or MRI coregistered to a CT simulation [31].

All of this may be challenging in an LMIC clinic. In situations where MRI is unavailable due to financial limitations one alternative is to use alternative imaging like CT with applicator in place and real-time transabdominal ultrasound (TAUS) or transrectal ultrasound (TRUS) to facilitate applicator placement and avoid uterine perforation [32]. TAUS and TRUS during brachytherapy have also been used to assist in target definition and treatment planning.

2.6 Emerging solutions for RT

In the quest to produce cost-effective healthcare solutions in radiotherapy there have been several innovations in the last decade to answer the call. These solutions are currently not available on the market hence they will be mentioned briefly in this section.

There is one proposed system that relies on kilovoltage (kV) beams. This is particularly attractive because the technology to produce kV beams is relatively simple (x-ray tubes). However, typically the superficial doses from a kV beam are too high to be used therapeutically. Recent proposals have advanced the idea that kV beams could spread out in a configuration that allows them to converge in a relatively small volume where the dose become large [33, 34]. Although there are challenges too with the dose rate and treatment planning there are currently commercial solutions being explored.

Additionally, there is the Cobalt-based IMRT system. Several studies have suggested that plan quality may be acceptable from such a system [35–38]. However, the dose rate from Cobalt is too low to be useable and the first-generation device from ViewRay Inc. used three Cobalt heads to deliver sufficient dose rate. However, more recently a proposal has been put forward to use compensators in combination with Cobalt beams which affords an advantage in terms of dose rate and in avoiding the mechanisms required for multileaf collimator [39]. More recent treatment planning simulations suggest this is feasible [40] and the design is moving into prototype production.

Finally, there is the fixed-beam and rotating patient system. Modern radiotherapy machines rotate the radiation beam around a patient. However, the fixed-beam system rotates the patient in a horizontal position relative to a fixed beam. These fixed-beam patient rotation radiotherapy machines show promise for reducing the size, surface area footprint, and shielding requirements compared to conventional rotating C-arm linacs. This system has been explored in support by Australian government grants dating back to 2012 and has since been commercialized by Nano-X technologies. For more details on this system see studies from Keal *et al* [41, 42]. Finally, a similar system has been proposed in which the patient is imaged and treated in a seated position (Leo Cancer Care Inc.). For a proton therapy clinic this would allow for a single fixed beamline and no gantry which would greatly reduce complexity and cost.

2.7 Conclusions

Cancer is emerging as a major challenge not just in high-income countries but in regions of the world where the burden of disease previously fell more to infectious diseases. A rapid expansion for cancer-care needs is expected in the coming decades. One of the key modalities in managing cancer is RT. However, the technical solutions to plan and deliver such therapy are complex and have emerged organically over the past 60 years. The technology that is presently available is not optimally suited for sustainable deployment in all environments and especially LMICs. RT technology needs to be rethought.

The principles for cost-effective solutions outlined here provide guidance for the path forward. These are consistent with the ‘CESER’ principles highlighted in chapter 1. Namely, *Cultural* awareness, *Ethical* integrity, *Social* responsibility and

Environmental Responsibility. In the context of RT technology this includes a consideration of upfront capital costs, the available infrastructure, the ability to service and maintain equipment, and the availability of appropriate staff to support the operations of this complex technology. While these goals may seem aspirational, technology solutions have already started to emerge which at least partially fulfil these design goals as outlined in the examples provided in this chapter. The decades ahead will see a rapid expansion in the global need for cancer care. It will be crucial to pursue technology development pathways that ensure safe and effective care for all patients.

References

[1] Bray F, Jemal A, Grey N, Ferlay J and Forman D 2012 Global cancer transitions according to the Human Development Index (2008–2030): a population-based study *Lancet Oncol.* **13** 790–801

[2] Board PDQATE 2002 *Financial Toxicity and Cancer Treatment (PDQ®): Health Professional Version. PDQ Cancer Information Summaries* (Bethesda, MD: National Cancer Institute (US))

[3] Atun R, Jaffray D A and Barton M B *et al* 2015 Expanding global access to radiotherapy *Lancet Oncol.* **16** 1153–86

[4] Radiation therapy in Africa: distribution and equipment *Radiother. Oncol.* 1999 **52** 79–83

[5] Abdel-Wahab M, Bourque J M and Pynda Y *et al* 2013 Status of radiotherapy resources in Africa: an International Atomic Energy Agency analysis *Lancet Oncol.* **14** e168–75

[6] International Atomic Energy Agency (IAEA) 2024 DIrectory of RAdiotherapy Centres (DIRAC) https://www.iaea.org/resources/databases/dirac

[7] Rick T J, Aagard M and Erwin E *et al* 2021 Barriers to cancer care in northern Tanzania: patient and health-system predictors for delayed presentation *JCO Global Oncol.* 1500–8

[8] Agodirin O, Olatoke S and Rahman G *et al* 2020 Presentation intervals and the impact of delay on breast cancer progression in a black African population *BMC Public Health* **20** 962

[9] Baskar R, Lee K A, Yeo R and Yeoh K W 2012 Cancer and radiation therapy: current advances and future directions *Int. J. Med. Sci.* **9** 193–9

[10] Balogh J M and Sutherland S E 1989 Osteoradionecrosis of the mandible: a review *J. Otolaryngol.* **18** 245–50

[11] Cooper J S, Fu K, Marks J and Silverman S 1995 Late effects of radiation therapy in the head and neck region *Int. J. Radiat. Oncol. Biol. Phys.* **31** 1141–64

[12] Harrison L B, Zelefsky M J and Pfister D G *et al* 1997 Detailed quality of life assessment in patients treated with primary radiotherapy for squamous cell cancer of the base of the tongue *Head Neck* **19** 169–75

[13] Veldeman L, Madani I, Hulstaert F, De Meerleer G, Mareel M, De and Neve W 2008 Evidence behind use of intensity-modulated radiotherapy: a systematic review of comparative clinical studies *Lancet Oncol.* **9** 367–75

[14] Yan M, Gouveia A G and Cury F L *et al* 2021 Practical considerations for prostate hypofractionation in the developing world *Nat. Rev. Urol.* **18** 669–85

[15] McCarroll R, Youssef B and Beadle B *et al* 2017 Model for estimating power and downtime effects on teletherapy units in low-resource settings *J. Glob. Oncol.* **3** 563–71

[16] Viegas C, Viswanathan A K and Lin M Y *et al* 2017 American Brachytherapy Society: brachytherapy treatment recommendations for locally advanced cervix cancer for low-income and middle-income countries *Brachytherapy* **16** 85–94
[17] Grover S, Longo J and Einck J *et al* 2017 The unique issues with brachytherapy in low- and middle-income countries *Semin. Radiat. Oncol.* **27** 136–42
[18] Ige T A, Hasford F and Tabakov A *et al* 2020 Medical physics development in Africa—status, education, challenges *Med. Phys. Int.* **8** 303–16
[19] Grout J 2006 Mistake proofing: changing designs to reduce error *Qual. Saf. Health Care* **15 Suppl 1** i44–9
[20] Norman D A 2002 *The Design of Everyday Things* (New York: Basic Books Inc.)
[21] Velarde A, Najera K D and Gay H *et al* 2020 Taking Guatemala from cobalt to IMRT: a tale of US agency collaboration with academic institutions and industry *Int. J. Radiat. Oncol. Biol. Phys.* **107** 867–72
[22] Cozzi L, Fogliata A and Thompson S *et al* 2018 Critical appraisal of the treatment planning performance of volumetric modulated arc therapy by means of a dual layer stacked multileaf collimator for head and neck, breast, and prostate *Technol. Cancer Res. Treat.* **17** 1533033818803882
[23] Weidlich G A, Bodduluri M, Achkire Y, Lee C and Adler J R 2019 Characterization of a novel 3 megavolt linear accelerator for dedicated intracranial stereotactic radiosurgery *Cureus.* **11** e4275
[24] Srivastava S P, Jani S S and Pinnaduwage D S *et al* 2021 Treatment planning system and beam data validation for the ZAP-X: a novel self-shielded stereotactic radiosurgery system *Med. Phys.* **48** 2494–510
[25] *Setting Up a Radiotherapy Programme* 2008 (Vienna: International Atomic Energy Agency) .
[26] Granero D, Pérez-Calatayud J and Ballester F 2007 Technical note: dosimetric study of a new Co-60 source used in brachytherapy *Med. Phys.* **34** 3485–8
[27] Ballester F, Granero D, Pérez-Calatayud J, Casal E, Agramunt S and Cases R 2005 Monte Carlo dosimetric study of the BEBIG Co-60 HDR source *Phys. Med. Biol.* **50** N309–16
[28] Tabrizi S H, Asl A K and Azma Z 2012 Monte Carlo derivation of AAPM TG-43 dosimetric parameters for GZP6 Co-60 HDR sources *Phys. Med.* **28** 153–60
[29] Dickler A, Kirk M C and Coon A *et al* 2008 A dosimetric comparison of Xoft Axxent electronic brachytherapy and iridium-192 high-dose-rate brachytherapy in the treatment of endometrial cancer *Brachytherapy.* **7** 351–4
[30] Pötter R, Tanderup K and Kirisits C *et al* 2018 The EMBRACE II study: the outcome and prospect of two decades of evolution within the GEC-ESTRO GYN working group and the EMBRACE studies *Clin. Transl. Radiat. Oncol.* **9** 48–60
[31] Fields E C and Weiss E 2016 A practical review of magnetic resonance imaging for the evaluation and management of cervical cancer *Radiat. Oncol.* **11** 15
[32] Mahantshetty U, Poetter R and Beriwal S *et al* 2021 IBS-GEC ESTRO-ABS recommendations for CT based contouring in image guided adaptive brachytherapy for cervical cancer *Radiother. Oncol.* **160** 273–84
[33] Paterno G, Marziani M and Camattari R *et al* 2016 Laue lens to focus an x-ray beam for radiation therapy *J. Appl. Crystallogr.* **49** 468–78
[34] Bazalova-Carter M, Weil M D, Breitkreutz D Y, Wilfley B P and Graves E E 2017 Feasibility of external beam radiation therapy to deep-seated targets with kilovoltage x-rays *Med. Phys.* **44** 597–607

[35] Adams E J and Warrington A P 2008 A comparison between cobalt and linear accelerator-based treatment plans for conformal and intensity-modulated radiotherapy *Br. J. Radiol.* **81** 304–10

[36] Choi C H, Park S Y and Kim J I *et al* 2017 Quality of tri-Co-60 MR-IGRT treatment plans in comparison with VMAT treatment plans for spine SABR *Br. J. Radiol.* **90** 20160652

[37] Wooten H O, Green O and Yang M 2015 Quality of IMRT treatment plans using a Co-60 MR image guidance radiation therapy system *Int. J. Radiat. Oncol. Biol. Phys.* **92** 771–8

[38] Schreiner L J, Joshi C P, Darko J, Kerr A, Salomons G and Dhanesar S 2009 The role of Cobalt-60 in modern radiation therapy: dose delivery and image guidance *J. Med. Phys.* **34** 133–6

[39] Van Schelt J, Smith D L and Fong N *et al* 2018 A ring-based compensator IMRT system optimized for low- and middle-income countries: design and treatment planning study *Med. Phys.* **45** 3275–86

[40] Oh K, Gronberg M P and Netherton T J 2023 A deep-learning-based dose verification tool utilizing fluence maps for a cobalt-60 compensator-based intensity-modulated radiation therapy system *Phys. Imag. Radiat. Oncol.* **26** 100440

[41] Liu P Z Y, Nguyen D T, Feain I, O'Brien R, Keall P J and Booth J T 2018 Technical note: real-time image-guided adaptive radiotherapy of a rigid target for a prototype fixed beam radiotherapy system *Med. Phys.* **45** 4660–6

[42] Feain I, Coleman L, Wallis H, Sokolov R, O'Brien R and Keall P 2017 Technical note: the design and function of a horizontal patient rotation system for the purposes of fixed-beam cancer radiotherapy *Med. Phys.* **44** 2490–502

IOP Publishing

Humanitarian Engineering for Global Oncology

Eric Ford

Chapter 3

Advanced information and communication technologies including artificial intelligence for global radiation oncology

Laurence E Court, Saurabh S Nair, Skylar S Gay and Joshua S Niedzielski

3.1 Introduction

The workflow for radiation therapy treatment planning involves multiple tasks that can benefit from advanced technologies such as artificial intelligence (AI) [1, 2]. This process typically begins with a computed tomography (CT) scan, followed by contouring and treatment planning, quality assurance tasks, and, finally, treatment administration (figure 3.1). Other crucial aspects of this work include training the clinical staff, ensuring the quality of imaging devices, providing decision support based on clinical trials, and implementing continuous quality improvement processes. These tasks are complex and require collaboration among highly trained professionals, but there is a global shortage of such personnel [3]. To address this shortage, innovative approaches are needed to improve workflow efficiencies and allow clinical teams to treat more patients without significantly increasing costs. AI and automation have been suggested as potential solutions, offering efficiency gains, improved consistency, and enhanced quality and safety [4]. AI can support various tasks within the workflow, including image quality assessment, autocontouring and planning, and quality assurance of generated contours and treatment plans. Additionally, AI-based approaches can potentially benefit staff training efforts.

Recent advancements in AI technology are rapidly becoming available as commercial solutions for clinical teams. Besides the traditional treatment planning system vendors, numerous companies now offer deep learning (DL)-based autocontouring, autoplanning, decision support solutions, and automated quality assurance. There are also initiatives, such as the Radiation Planning Assistant at the University of Texas MD Anderson Cancer Center, specifically designed to provide these tools at minimal or no cost to clinics in low- and middle-income countries (LMICs) [5].

doi:10.1088/978-0-7503-3751-9ch3

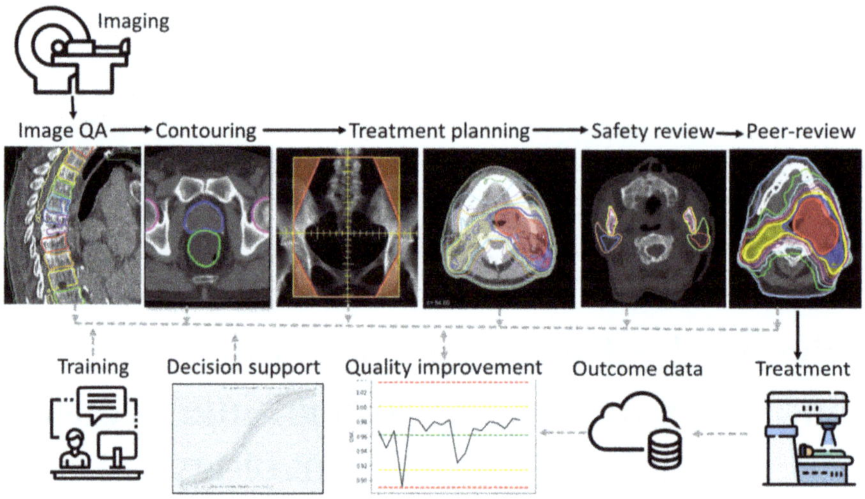

Figure 3.1. Overview of the workflow and associated tasks for radiation therapy planning.

This chapter aims to explore the application of advanced technologies, including AI, in global radiation oncology. The primary objective is to improve access to high-quality radiotherapy worldwide, particularly in LMICs.

3.2 Treatment decision support

Once it has been determined that the patient will benefit from radiotherapy, the oncologist must make decisions about the type of treatment, including the prescribed dose, based on a multitude of factors including individual patient anatomy, current medical condition or status, previous radiation treatment, whether they received induction chemotherapy, and other factors. Treatment decision support for radiotherapy is challenging due to the heterogeneity of patient populations and personalized cancer medicine has arisen in recent years to optimize therapeutic strategies for the variety of patient factors and treatment option that currently exist [6]. Moreover, AI is well suited to address the complex data associated with personalized cancer care to identify optimal treatment approaches for unique individual patients. There is an increasing amount of literature on applications of AI to treatment decision support, helping the clinician make these decisions based on quantitative data and predictions of patient outcomes [2, 7, 8]. Eventually these are likely to be useful globally, including situations where there are limited resources for additional treatment review (e.g. peer review of a proposed treatment plan by experienced colleagues), although consideration of local populations will be important.

Ultimately, treatment decision support uses past clinical experience, typically in the form of patient outcomes, coupled with baseline patient data from retrospective cohorts to determine optimal treatment strategies for future prospective patients [9]. Important methodologies in the treatment decision support process include outcome prediction modeling (OPM), multiomics, and knowledge-based planning (KBP).

Each of these components of treatment decision support will be discussed in the following subsections.

3.2.1 OPM of toxicity and tumor control

OPM is used to determine tumor response and/or treatment-related toxicities using a statistical model based on baseline patient features (i.e. disease stage, radiation dose, performance status, etc). While OPM is central to determining if a potential treatment strategy is viable for a given patient, in terms of both efficacy (i.e. tumor control probability and survival) and safety (i.e. normal tissue complication probability and toxicity), early modeling methods were limited in both scope and application as a result of poor model generalizability, tendency to overfit when large amounts of covariates were present, and an inability to handle high dimensional data [10]. AI-based implementations of OPM using machine learning (ML) and DL have not only greatly increased model performance in terms of predicting treatment response, but these data-driven strategies represent a paradigm shift from using model-based approaches of the past [11, 12]. Indeed, ML- and DL-based OPM are able to take large amounts of patient data in the form of clinical factors, imaging data, and genomic data, among others, to predict treatment outcomes with high performance.

The main clinical utility of AI-based OPM includes treatment decision support of candidate radiation treatment plans. For example, the ability to predict the treatment response of involved lymph nodes for patients with non-small cell lung cancer could inform the decision to continue treating with chemoradiotherapy or to proceed to surgery [13]. Outcome prediction models are already in use to help identify patients who may benefit most from proton therapy in comparison with photon therapy [14, 15]. A similar approach could potentially be applied to compare different photon-based treatment approaches (e.g. conventional conformal radiotherapy versus intensity-modulated radiation therapy (IMRT)), allowing the clinical teams to personalize their approach based on both the patients' anatomy and the available resources. In the low-resource setting, AI-based OPM could be leveraged to streamline clinical decision-making for each patient, by identifying the optimal treatment strategy for every unique patient, thereby allowing for potentially improved patient outcomes, without putting additional pressure on limited clinical resources.

3.2.2 Multiomics for treatment decision support

Multimomics encompasses a multitude of large data types that can be potentially leveraged for personalized medicine including genomic, proteomic, metabolomic, transcriptomic, radiogenomic, radiomic, and dosiomic [16]. Within the context of radiation oncology, and particularly the low-resource setting, radiomic and dosiomic data are of particular interest because these data require no extra overhead or test kits to extract, other than an appropriate software platform, many of which are freely available [17–19]. This subsection will focus on dosiomic- and radiomic-based multiomic data. Ultimately radiomic and dosimoic data have an undeniable synergy with OPM, as they can represent highly informative input data to improve OPM for treatment decision support.

Radiomics is the extraction of quantitative imaging features for decision support [9]. Typical imaging modalities used in radiomic analyses include CT, positron emission tomography (PET), and magnetic resonance imaging (MRI). One of the driving factors in the popularity of radiomic analysis within radiation oncology is that CT, MRI, and PET are commonly used in the diagnostic workup of many patients. Moreover, CT is the standard-of-care approach for radiation treatment planning of external beam therapy. Therefore, there is no additional imaging study needed beyond CT for most radiation oncology patients, thereby providing a 'free' source of radiomic data for treatment decision support. Radiomic features include shape, texture, and wavelet features, which are all based on mathematical formulations that mine the three-dimensional (3D) voxelized information in MRI, PET, and CT images [9]. Moreover, radiomic data can provide unique representation of data that reveals phenotypic and underlying pathophysiology representations that cannot be discerned from the anatomical images alone [20, 21]. This allows radiomic data to be potential imaging biomarkers of treatment response, for either tumor response or normal tissue toxicity, and thus aid in treatment decision support by stratifying patients based on baseline radiomic characteristics. Typical applications of radiomics include the patient-specific prediction of prognosis [20], treatment response [22], and toxicity [23] after radiation therapy.

The field of dosiomics is a recent development in research, in which many of the same principles applied to radiomics are extrapolated to radiation dose data [24]. As external beam radiation therapy treatment plans are quantified by 3D dose data, these can represent an 'image' that can have the same radiomic-based mathematical features extracted and used for treatment decision support. While not an inherent phenotypic trait of a given patient, extracted dosiomic features represent unique representations of the radiation dose delivered to the patient, and can therefore provide additional insight into treatment response in a retrospective analysis [25]. Central to dosiomic analyses is that the 3D dose distribution is examined for relationships to the outcome of interest; this contrasts with traditional methods of analyzing radiation dose which includes dose–volume histogram (DVH) metrics, where the 3D radiation distribution is reduced to a two-dimensional representation of dose as a function of volume, thereby removing the spatial connection of neighboring dose voxels in the distribution. It is feasible that this reduction strategy obscures spatial relationships of dose response. For example, Rossi *et al* showed that dosiomic features improved OPM of gastrointestinal and genitourinary toxicity for prostate patients, when compared to using DVH metrics alone [26]. Similarly, Jin *et al* used a combined radiomic–dosiomic model to predict patient response to radiotherapy for esophageal cancer [27]. Since dosiomics analyses extract features from the entire 3D dose as an image, individual dosiomic features can quantify associations within the dose data that traditional DVH methods cannot.

3.2.3 Decision support using knowledge-based treatment planning

With recent advances in AI methods, KBP for radiotherapy treatment has emerged as a critical tool for clinical resource optimization [28]. Early implementations of

KBP for radiotherapy included predicting DVHs and specific dose metrics for a given region of interest. However, with the maturation of AI-based methods, particularly convolutional neural networks and generative adversarial networks (GANs), KBP can accurately predict the 3D dose distribution of a proposed treatment approach for a given patient [29, 30]. While KBP can be used to help lower resource burden for a clinic by reducing the manpower necessary to create radiotherapy treatment plans [30], another critical application of KBP is treatment decision support.

One possible application of KBP for treatment decision support includes determining if a proposed prescription and fractionation schedule is appropriate for a given patient. KBP planning, when coupled with OPM, can also be used to compare several treatment approaches based on modality [31] thereby determining which precise radiation treatment is ideal for a specific patient. Indeed, KBP can allow for truly personalized cancer care by using a data-driven approach to select a patient's optimal treatment strategy. KBP can also be used to determine when treatment adaptation might be necessary due to changes in a patient's tumor or normal tissues [32, 33]. In the low-resource setting, validated AI-based KBP models can potentially be utilized to not only reduce clinical burden by streamlining the radiotherapy planning process, but also allow for the patient-tailored treatment strategies. In this context, widespread adoption of KBP approaches in low-resource clinics could improve global radiotherapy outcomes by aiding clinicians in selecting the optimal treatment approach for every patient.

3.3 Automated contouring

Contouring is the process of delineating the treatment targets and normal tissues [34]. Treatment planners or therapists often contour the normal tissues, while radiation oncologists typically take responsibility for contouring the treatment targets. These contours guide the treatment planning process, enabling planners to optimize the dose distribution for the targets while considering the dose tolerances of surrounding normal tissues.

For instance, contouring all necessary structures in the head and neck or abdomen can take several hours per patient. Interuser variations in contouring are noticeable, especially for target structures where different clinicians may have varying interpretations of what needs to be treated, with guidelines and local consensus (peer review) helping maintain quality. Although earlier automated contouring tools were often unreliable, DL has significantly improved the automatic contouring of normal tissues [35]. Some examples are shown in figure 3.2. The use of DL for autocontouring of targets has also been the subject of much research, although this has yet to make its way into routine clinical practice. Thus, AI-based contouring has the potential to significant improve clinical workflows, reducing the time taken to process the contours for each patient and helping the clinical teams scale their efforts to treat more patients.

There are some important considerations when deploying AI-based contouring into clinical practice. While AI-based autocontouring is significantly better than

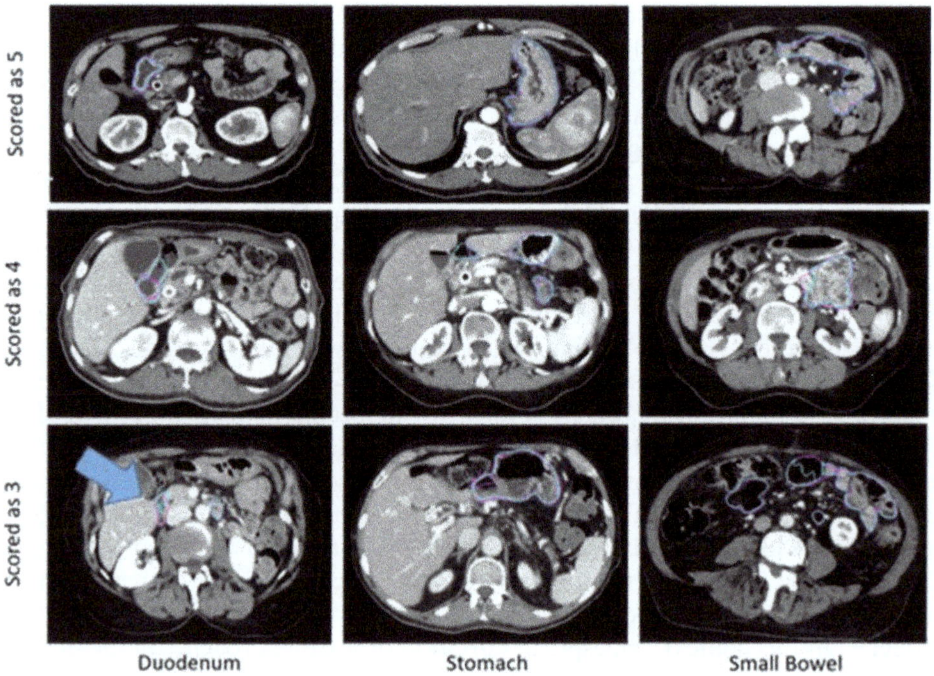

Figure 3.2. Representative contours of organs autocontoured in the abdomen, scored on a Likert scale as 5 (use as is), 4 (minor edits, not necessary for treatment), and 3 (minor edits, necessary for treatment) by physicians. The ground truth contours are shown as purple in all images. The automatically generated contours are shown as cyan in all images. The arrow indicates a segment of undercontoured duodenum that required minor edits. Reprinted by permission from Springer Nature [36], Copyright (2022).

earlier approaches, it is not perfect and can sometimes make unexpected errors when presented with situations that it did not experience in the training set. For example, if an abdomen autocontouring model that was only trained with abdomen cases is presented with an extended dataset that was not present in the training cases, it may do something very strange like contour the liver in the brain. Clinicians need to be aware and spend time reviewing all patients. Moreover, the AI models do not necessarily have access to the same data that clinicians do—for example, clinicians are accustomed to having access to additional information, such as physical exam, endoscopic data, and other images, that are often not available to the autocontouring algorithm [37].

3.4 Automated radiotherapy planning

The quality of the treatment plan defines the dose distribution that will be delivered to the patient, including how well the therapeutic radiation dose is delivered to the prescribed targets (related to tumor control) and the dose that nearby normal tissues receive (related to treatment-related toxicity). Thus, plan quality is directly related to patient outcomes. Furthermore, errors in the treatment planning process can, if undetected, result in harm to the patient, especially because the same radiation plan is typically used to treat the patient for their entire course of radiation treatment which can be 7 weeks long.

The treatment planner designs the optimal plan using specialized software called a Treatment Planning System, adjusting the beam arrangement, field shape, and radiation intensity. This can be a difficult, complex, and time-consuming process. Although many steps are heavily automated, such as the determination of the intensity modulation in inverse-planned IMRT, there are still a huge number of manual processes and decisions that must be made throughout the process. This means that a single plan for one patient can take a treatment planner several hours [38]. The extensive manual input to a treatment plan also means that quality can be quite variable. Different radiation oncologists have different expectations regarding the compromises that are made in the planning process to create a clinically acceptable plan. There is also variation in treatment planners, with some giving much higher-quality plans than others [39].

These challenges can be exacerbated in centers with insufficient staffing, where there may not be sufficient time to spend iterating complex plans. AI and automation have been suggested as a potential route to help these centers scale their efforts, potentially also helping reduce the gaps in quality.

3.4.1 AI for simple treatment planning techniques

Radiotherapy can be planned and delivered in several different ways, with the approach determined by local clinical practices, specific to the current patient, and available resources (equipment and staff). Simple treatment plans may involve a limited number of treatment fields, without complex intensity modulation. They can treatment patients effectively and be planned in less time and each treatment is typically faster than the complex treatments. The use of AI to automate planning for these simple treatments can potentially free up planners' time such that they can spend more time working on more complex cases. Examples of simple planning approaches that have been successfully automated include treatments of whole brain [40], rectal cancer [41], breast cancer (intact [42] and postmastectomy [43]), and cervical cancer [44]. One challenge in developing automated approaches for simple plans is interuser variability, with clinicians using quite different field shapes to achieve the same goal [37]. This means that it can be useful to include some level of customization, where specifics of the treatment approach can be fine-tuned to suite local clinical practice. One example of this for whole-brain treatments is shown in figure 3.3 [40].

Even simple plans can be extremely time-consuming for treatment planners to generate. Pediatric craniospinal irradiation therapy, for example, involves careful matching of adjacent radiation fields and regular shifting of these match-lines to mitigate potential in homogeneous dose distributions. Even the simplest craniospinal irradiation (CSI) plan involves nine matched fields, although this can easily double when subfields are introduced to reduce avoidable radiation hotspots. Details of the field arrangements are also dependent on the patient, adding additional

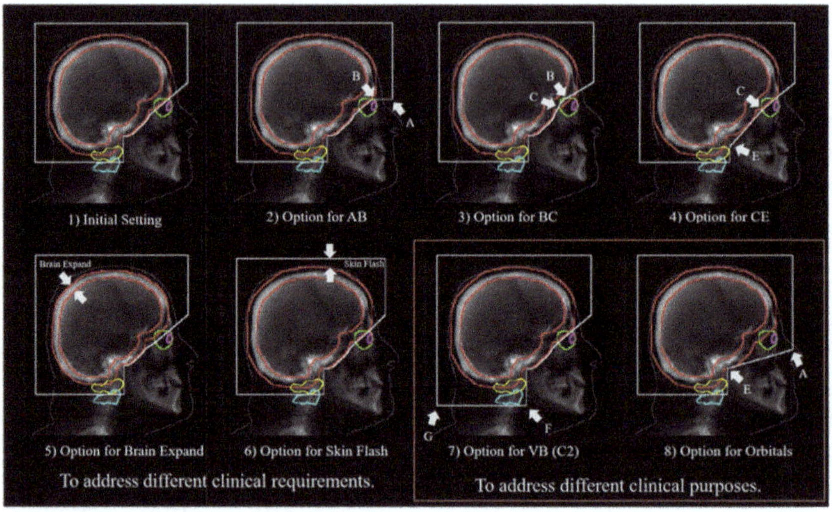

Figure 3.3. Generated field apertures based on different options. Options 1–6 are configured to address different clinical preferences, and options 7 and 8 are configured to address different clinical purposes. From Xiao *et al* [40] John Wiley & Sons.

complexity. Hernandez *et al* have shown that this entire process can be fully automated, with high levels of acceptability, potentially saving significant planning time [45].

3.4.2 AI for complex treatment planning techniques

IMRT plans provide complex dose distributions that are highly conformal to the treatment targets and minimize the radiation dose to nearby normal tissues. The usual planning process involves setting optimization goals in the Treatment Planning System (TPS) and then running an optimization process to find the x-ray fluence or multileaf collimator sequence that gives a dose distribution that best meets the preset goals. The main challenge for the treatment planner involves setting reasonable objectives and priorities to achieve an acceptable treatment plan, as these can vary between patients as the geometric relationship between targets and normal tissues varies. This is typically a process of trial and error, either changing the objectives on the fly (as the system optimizes the plan), or iteratively (by adjusting and reoptimizing the plan after reviewing an optimized plan). This can be very time-consuming.

KBP, where an AI model is trained to predict what the final treatment plan will look like (typically as a DVH), has been reported to address this challenge, helping treatment planners achieve plans that have reduced normal tissue doses and are also more consistent from patient to patient [46, 47]. These tools are also ideal when automating the treatment planning process. Rhee [48] and Olanrewaju [49] have both reported very high clinical acceptability rates for automated volumetric modulated arc therapy planning based on KBP models (cervical cancer and head and neck cancer, respectively).

3.5 Plan approval and quality assurance

Once a radiotherapy treatment plan is complete, it must be reviewed and approved by the attending radiation oncologist. It also passes through several layers of quality assurance checks to ensure that the plan is appropriate, safe, and optimal for patient treatment. These include physics pretreatment plan review and peer review of the treatment plan. The physics review, which has been shown to be the most effective plan quality assurance (QA) steps, reviews many aspects of the treatment plan including dosimetry and plan parameters [50]. The peer review process involves review of the proposed treatment by clinical peers (i.e. other radiation oncologists), to ensure that it is appropriate. Both of these processes provide important contributions in driving the quality and safety of radiotherapy treatments. They are, however, time-consuming and do not catch all errors. Physics checks have been shown to catch around half of planning errors [50]. The effectiveness of peer review by radiation oncologists is similar [51]. Thus, AI may play an important role in supporting these efforts, reducing the risk of errors reaching the patient by scaling up the number of errors it can detect.

It should also be noted that the use of AI to automate contouring and planning may also introduce unexpected risks to the process [52, 53]. These include automation bias, where users are overly reliant on the quality of the AI models. Increasing the role of automation may also reduce opportunities for the users to notice and rectify errors. For these reasons, the physician and physics reviews have additional importance when AI is introduced into the clinical workplace. It can be advantageous to develop additional manual QA processes based on the new errors that automation may be expected to introduce to the clinical workflow [54].

AI and automation can supplement these review processes in several ways. AAPM's Task Group 275 identified several tasks that could be improved with automation [50]. It has also been shown that adding automated quality assurance into the physics review process can improve the likelihood of errors being caught, and reduce the overall risk profile of radiotherapy treatment planning [52]. This includes approaches such as using AI to determine the expected field shapes and then comparing these predictions with the treatment plan and flagging any unusual differences. AI can also support the physician review of treatment plans. Gronberg *et al*, for example, showed that DL predictions of dose distributions can be used to flag treatment plans that can benefit from additional optimization to further reduce normal tissue doses [55]. The use of AI for decision support (see earlier section) will also aid in ensuring that each patient receives the optimal treatment.

3.6 Treatment delivery

Advanced technologies, especially AI, are likely to touch all aspects of radiation therapy delivery, although the research in this area is less advanced than the autocontouring and planning work described above. Here we will describe three examples: (1) use of AI to explore the health of the treatment devices, predict breakdown, and therefore minimize unscheduled downtime; (2) use of AI to predict deliverability of specific patient plans; and (3) applications of AI for image registration to improve the workflow and precision of image-guided radiation therapy treatment delivery.

Radiotherapy treatments use large, complex treatment devices called medical linear accelerators (linacs). Each patient receives many repeat treatments, with most receiving daily treatments over several weeks, and any delay in completing treatment can have a negative impact on patient outcomes. It is important, therefore, to minimize linac downtime, and to perform repairs in as short a time as possible. This is a particular challenge in many clinics in LMICs, as the service personnel and spare parts are not local, and this can result in significant delays. One way that has been proposed to try to mitigate these issues is the use of AI to monitor machine parameters or daily QA results, and predict breakdowns. Overall, the research in this area is promising, and we can expect more developments in this area, hopefully leading to reduced unscheduled downtime for clinics in relatively removed regions [56, 57].

The planning and delivery of complex dose distributions involve many radiation fields which have complex, often small, shapes that are created using very precise movements of the collimators in the linac. The complexity of these dose calculations and delivery, as well as incidences of incorrect patient treatment, have led to widespread adoption of patient-specific experiments to verify the dose distribution prior to treatment. These experiments can be time-consuming, and require the appropriate resources (equipment and personnel). For this reason, many groups have investigated the use of AI to predict the passing rate for individual patient plans. Although these models have yet to be integrated into the clinical workflow, they have consistently shown that AI can perform this role reliably. It seems likely that they will gradually make their way into our clinical processes, probably starting with triaging the plans with the lowest predicted passing rates for additional quality assurance, and ultimately improving the efficiency of our workflows and improving the ability of clinics with limited resources to offer these complex treatments [58].

Image-guided radiation therapy describes the registration of images taken immediately prior to treatment with the treatment planning images, followed by the correction of the patient's position, thus ensuring precise treatment of the target and reducing unnecessary radiation doses to the surrounding normal tissues. Adaptive radiation therapy starts with a similar process, but involves adapting the treatment plan to the daily patient's geometry. This is much less common, but increasing, and may be advantageous when significant changes such as tumor shrinkage or internal motion (such as due to badder filling), mean that the original plan is no longer optimal. AI-based rigid or deformable registration is the subject of increased research, and it seems likely that it will lead to improved speed and precision [59].

3.7 Patient follow-up

Patient follow-up can be a significant challenge, as many patients do not return to the treating clinic because of cost, distance, time, or other reasons. A lack of clinical infrastructure, such as no available central Electronic Medical Records (EMR) system, can also complicate the efficient collection and warehousing of follow-up data. It is likely that, by leveraging AI technologies, healthcare providers can enhance the efficiency and effectiveness of patient follow-up monitoring, improve patient outcomes by accurately capturing and analyzing treatment efficacy, and optimize resource allocation by identifying areas of need or deficit. These include risk stratification and automatically flagging patients who may require additional attention or intervention (enabling healthcare providers to prioritize follow-up care), remote patient monitoring (e.g. wearable monitoring devices) [60], and patient engagement. The application of these AI-driven optimization strategies to local clinics that are resource deficient will have to be carefully evaluated on a case-by-case basis.

AI-powered chatbots and virtual assistants may also be utilized to engage patients in post-treatment follow-up monitoring by providing personalized education, answering questions, and addressing concerns. These AI-driven tools can help patients better understand their condition, treatment, and self-care strategies, empowering them to actively participate in their own follow-up care and make informed decisions. Similar concepts can also be applied to helping the patients understand the radiotherapy treatment process and thereby discern what they will encounter and can expect during treatment (i.e. pretreatment education) [61, 62]. Indeed, this can help improve patient compliance with radiotherapy procedures and provide superior patient experience [63, 64].

3.8 Risk and challenges to clinical implementation

Introducing advanced technologies into clinical practice can introduce new, sometimes unexpected risks. Although it might be assumed the main risks come from errors/faults in the technology itself (e.g. incorrectly contoured structures), this is not necessarily the case. It is therefore advisable to perform extensive validation, usability testing, risk assessment, and hazard testing.

Nealon *et al* performed a failure modes and effects analysis of an automated contouring and planning workflow, and found that operator error and off-label use (where the operator uses the software for something for which it was not intended or tested) were likely to occur more frequently than software error [53]. They estimated that the most common high-risk failure modes were operator error, automation bias, and software error. An example of automation bias is when the user over-relies on the automated processes and, therefore, does not check them as thoroughly as they might for manual contours or plans.

Operator error is generally caused by incorrect data entry, emphasizing the importance of careful design and validation of the user interfaces, as well as the automated algorithms. Examples include restricting data entry (prescribed dose, number of fractions), and careful presentation of all information such that errors are more easily identified. It should be remembered that design endpoints should be focused on safety, as well as usability, as users may report that a system is easy to use and they are confident in their abilities while, at the same time, hazard testing may show that the users are not identifying system errors.

The need for manual contour and plan review of automatically generated contours and plans has been emphasized in several studies [52, 58]. This is partly in response to the risk of automation bias, but also because having fewer people involved in the contouring/planning process means that any errors are more likely to make their way to the end of the process without being detected. Nealon went one step further and showed that plan review checklists that are designed with automated processes in mind, can be useful in improving the likelihood of a manual plan check catching errors in the automated processes [54]. Specifically, they intentionally introduced errors and then tested the ability of physicists to catch these errors with and without the new checklist, and found that error detection was improved by 17% with the checklist. Automated quality assurance (section plan approval and quality assurance) may also likely play a role in reducing risk when automated solutions are introduced to the clinic.

In addition to risk, there are many other challenges to the successful deployment of an AI-based tool to clinics in LMICs, all of which should be considered when designing, building, and deploying a new system. McGinnis surveyed radiotherapy providers in South Africa, Botswana, Tanzania, and Guatemala, and identified the lack of reliable internet and possible costs as the main barriers to using the Radiation Planning Assistant, a web-based tool. They identified increased patient throughput (for planning), and planning consistency as the main advantages.

3.9 Scaled solutions

AI is increasingly becoming integrated into radiotherapy vendors' offerings. RaySearch Laboratories (Stockholm, Sweden) has implemented ML-based prostate plan design [65] and DL segmentation into the RayStation Treatment Planning System versions 8B and 11B, respectively. Varian Medical Systems (Palo Alto, USA) provides knowledge-based RapidPlan DVH predictions of high-quality plans, and integrates with AI-Rad Companion Organs RT (Siemens Healthineers, Erlangen, Germany) for automated contouring of organs at risk (OARs).

Additionally, the Varian Ethos linac provides an AI-based, on-couch adaptive radiotherapy workflow. Many vendors also offer AI as the cornerstone of their services. As examples in this rapidly growing space, AiContour (Linking MED, Beijing, China) and Limbus AI (Regina, Canada) offer automated contouring of OARs and target volumes from medical images, while Radformation (New York City, USA) adds automated planning and quality assurance as well.

The Radiation Planning Assistant [5], a 501(k) cleared but not yet marketed product, is a cloud-based, fully automated radiotherapy planning system. Developed at MD Anderson Cancer Center, it will provide no- or low-cost, high-quality radiotherapy plans with a focus on the needs of LMICs.

Finally, there are various freewares available which utilize AI and are often released by enthusiasts. Examples of these include TotalSegmentator (https://github.com/wasserth/TotalSegmentator), a tool that automatically contours 104 structures from CT images, is based on the popular nnUNet segmentation architecture [66], and has been integrated as a plugin (https://github.com/lassoan/SlicerTotalSegmentator) to the popular 3D Slicer [67] medical image visualization tool. MONAI Label provides an adaptable framework for rapidly annotating regions of interest or OARs from medical images.

The introduction of DL has significantly reduced the technical barriers to entry for vendors, startup companies, and research groups, and we hope that this will result in reduced cost for the end users who will not otherwise benefit from these technological advances.

3.10 AI and the global radiotherapy workforce

AI will inevitably change the roles of professionals like radiation oncologists, medical physicists and dosimetrists. Many tasks, such as contouring and planning, will involve extensive use of automation, so these clinical team members will have less direct involvement in these tasks. It seems likely, however, that manual plan review will remain a vital task for the clinical team [53]. It will become increasingly important that they are trained in how to review contours and plans created by AI. The role of these team members will also likely evolve so that less emphasis is placed on manual tasks, and more on development and the evaluation of radiation treatments [68].

AI and other advanced technologies are also likely to play an increasing role in how we train our clinical teams. Practical training is vital for our radiation oncologists. However, for example, the median number of head and neck cancer (HNC) cases (multiple sites, including oral cavity, oropharynx, nasopharynx, etc) treated during residency is only 53 (range: 6–143); almost 25% of residents see fewer than 10 HNC patients per year [69]. Overall, this may mean that many practicing radiation oncologists have insufficient training and continuing experience in all aspects of treating patients with HNC. It seems likely that AI will eventually play a role in training, perhaps using GANs or similar models to artificially create cases for review and training, as has become common for other radiology applications [70].

Overall, although we cannot predict exactly how AI will affect these roles, the potential benefits to the patient are enormous, and it seems likely that AI will help us scale our efforts to treat more patients.

3.11 Infrastructure

The application of advanced technologies requires the appropriate infrastructure. A survey of radiation therapy providers in sub-Saharan Africa and Central America by McGinnis *et al* found that poor internet infrastructure is one of the most anticipated hurdles to the effective use of web-based tools such as automated contouring and planning [71]. As per a study conducted in 2021, only around 25% of people residing in LMICs have access to the internet [72]. Having access to good-quality internet is imperative as it enables healthcare professionals to obtain and share medical information, monitor diseases, and collaborate and interact with other professionals throughout the world [73]. Having access to high-speed internet helps to address another major roadblock to providing high-class radiotherapy—shortage of human resources. High patient volumes and lack of trained personnel are significant causes of concern in LMICs and having a good connectivity with the world will facilitate more online training programs and collaborations which in-turn will better equip LMICs to overcome this daunting challenge [74]. With the goal of providing internet connectivity to the rural world, the idea of using low earth orbit satellites has become increasingly popular in the past few years. Some of the biggest names developing this technology—Space X's Starlink, Amazon's Kuiper Project, along with others—have pledged to bring fast, affordable internet to unserved and underserved countries around the world this coming decade.

Another significant infrastructure barrier to providing high-quality radiotherapy in LMICs is the lack of implementation of EMR systems in healthcare facilities. A survey conducted by the World Health Organization states only 15% of low-income countries have adopted electronic record systems in health institutions [75]. EMRs are much more beneficial than paper medical records as they improve efficiency and patient safety, reduce medical errors, and improve administrative processes like care management and care delivery and hence should be introduced into all healthcare systems in LMICs [76]. They also provide the basis of the infrastructure for research into patient outcomes.

3.12 Conclusions

We are in an exciting time, where advanced technologies are starting to make a tangible difference to workflows and how patients are treated, potentially improving access to high-quality radiotherapy across the world. Potential advantages include scaling the efforts of clinical teams so that they can treat more patients, as well as improving the overall quality and consistency of contours and treatment plans.

Acknowledgments

We would like to thank Barbara Marquez for reviewing this chapter. Our team would also like to thank our funding partners, NCI, CPRIT, Wellcome Trust, and the Fund for Innovation in Cancer Informatics.

References

[1] Netherton T J, Cardenas C E, Rhee D J, Court L E and Beadle B M 2020 The emergence of artificial intelligence within radiation oncology treatment planning *Oncology* **99** 124–34

[2] Huynh E, Hosny A, Guthier C, Bitterman D S, Petit S F and Haas-Kogan D A *et al* 2020 Artificial intelligence in radiation oncology *Nat. Rev. Clin. Oncol.* **17** 771–81

[3] Datta N R, Samiei M and Bodis S 2014 Radiation therapy infrastructure and human resources in low- and middle-income countries: present status and projections for 2020 *Int. J. Radiat. Oncol., Biol., Phys.* **89** 448–57

[4] Atun R, Jaffray D A, Barton M B, Bray F, Baumann M and Vikram B *et al* 2015 Expanding global access to radiotherapy *Lancet Oncol.* **16** 1153–86

[5] Court L E, Kisling K, McCarroll R, Zhang L, Yang J and Simonds H *et al* 2018 Radiation planning assistant—a streamlined, fully automated radiotherapy treatment planning system *J. Vis. Exp.* **134** 57411

[6] Abernethy A P, Etheredge L M, Ganz P A, Wallace P, German R R and Neti C *et al* 2010 Rapid-learning system for cancer care *J. Clin. Oncol.* **28** 4268–74

[7] Valdes G, Simone C B, Chen J, Lin A, Yom S S and Pattison A J *et al* 2017 Clinical decision support of radiotherapy treatment planning: a data-driven machine learning strategy for patient-specific dosimetric decision making *Radiother. Oncol.* **125** 392–7

[8] Niraula D, Sun W, Jin J, Dinov I D, Cuneo K and Jamaluddin J *et al* 2023 A clinical decision support system for AI-assisted decision-making in response-adaptive radiotherapy (ARCliDS) *Sci. Rep.* **13** 5279

[9] Lambin P, Zindler J, Vanneste B G L, De Voorde L V, Eekers D and Compter I *et al* 2017 Decision support systems for personalized and participative radiation oncology *Adv. Drug Deliv. Rev.* **109** 131–53

[10] Luo Y, Chen S and Valdes G 2020 Machine learning for radiation outcome modeling and prediction *Med. Phys.* **47** e178–84

[11] Breiman L 2001 Statistical modeling: the two cultures (with comments and a rejoinder by the author) *Stat. Sci.* **16** 199–231, 33

[12] Cui S, Hope A, Dilling T J, Dawson L A, Ten Haken R and El Naqa I 2022 Artificial intelligence for outcome modeling in radiotherapy *Semin. Radiat. Oncol.* **32** 351–64

[13] Kann B H, Aneja S, Loganadane G V, Kelly J R, Smith S M and Decker R H *et al* 2018 Pretreatment identification of head and neck cancer nodal metastasis and extranodal extension using deep learning neural networks *Sci. Rep.* **8** 14036

[14] Langendijk J A, Hoebers F J P, de Jong M A, Doornaert P, Terhaard C H J and Steenbakkers R J H M *et al* 2021 National protocol for model-based selection for proton therapy in head and neck cancer *Int. J. Part. Ther.* **8** 354–65

[15] Niedzielski J S, Yang J, Mohan R, Titt U, Mirkovic D and Stingo F *et al* 2017 Differences in normal tissue response in the esophagus between proton and photon radiation therapy for non-small cell lung cancer using *in vivo* imaging biomarkers *Int. J. Radiat. Oncol. Biol. Phys.* **99** 1013–20

[16] El Naqa I 2014 Biomedical informatics and panomics for evidence-based radiation therapy *WIREs Data Min. Knowl. Discov.* **4** 327–40

[17] Zhang L, Fried D V, Fave X J, Hunter L A, Yang J and Court L E 2015 IBEX: an open infrastructure software platform to facilitate collaborative work in radiomics *Med. Phys.* **42** 1341–53

[18] Gillies R J, Kinahan P E and Hricak H 2016 Radiomics: images are more than pictures, they are data *Radiology* **278** 563–77

[19] Placidi L, Gioscio E, Garibaldi C, Rancati T, Fanizzi A and Maestri D *et al* 2021 A multicentre evaluation of dosiomics features reproducibility, stability and sensitivity *Cancers* **13** 3835

[20] Aerts H J, Velazquez E R, Leijenaar R T, Parmar C, Grossmann P and Carvalho S *et al* 2014 Decoding tumour phenotype by noninvasive imaging using a quantitative radiomics approach *Nat. Commun.* **5** 4006

[21] Tomaszewski M R and Gillies R J 2021 The biological meaning of radiomic features *Radiology* **298** 505–16

[22] Liu Z, Zhang X Y, Shi Y J, Wang L, Zhu H T and Tang Z *et al* 2017 Radiomics analysis for evaluation of pathological complete response to neoadjuvant chemoradiotherapy in locally advanced rectal cancer *Clin. Cancer Res.* **23** 7253–62

[23] Sheikh K, Lee S H, Cheng Z, Lakshminarayanan P, Peng L and Han P *et al* 2019 Predicting acute radiation induced xerostomia in head and neck cancer using MR and CT Radiomics of parotid and submandibular glands *Radiat. Oncol.* **14** 131

[24] Liang B, Yan H, Tian Y, Chen X, Yan L and Zhang T *et al* 2019 Dosiomics: extracting 3D spatial features from dose distribution to predict incidence of radiation pneumonitis *Front. Oncol.* **9** 269

[25] Lam S-K, Zhang Y, Zhang J, Li B, Sun J-C and Liu C Y-T *et al* 2022 Multi-organ omics-based prediction for adaptive radiation therapy eligibility in nasopharyngeal carcinoma patients undergoing concurrent chemoradiotherapy *Front. Oncol.* **11** 792024

[26] Rossi L, Bijman R, Schillemans W, Aluwini S, Cavedon C and Witte M *et al* 2018 Texture analysis of 3D dose distributions for predictive modelling of toxicity rates in radiotherapy *Radiother. Oncol.* **129** 548–3

[27] Jin X, Zheng X, Chen D, Jin J, Zhu G and Deng X *et al* 2019 Prediction of response after chemoradiation for esophageal cancer using a combination of dosimetry and CT radiomics *Eur. Radiol.* **29** 6080–8

[28] Ge Y and Wu Q J 2019 Knowledge-based planning for intensity-modulated radiation therapy: a review of data-driven approaches *Med. Phys.* **46** 2760–75

[29] Gronberg M P, Gay S S, Netherton T J, Rhee D J, Court L E and Cardenas C E 2021 Technical Note: dose prediction for head and neck radiotherapy using a three-dimensional dense dilated U-net architecture *Med. Phys.* **48** 5567–73

[30] Zhang J, Wang C, Sheng Y, Palta M, Czito B and Willett C *et al* 2021 An interpretable planning bot for pancreas stereotactic body radiation therapy *Int. J. Radiat. Oncol. Biol. Phys.* **109** 1076–85

[31] Cheng Q, Roelofs E, Ramaekers B L T, Eekers D, van Soest J and Lustberg T *et al* 2016 Development and evaluation of an online three-level proton vs photon decision support prototype for head and neck cancer - comparison of dose, toxicity and cost-effectiveness *Radiother. Oncol.* **118** 281–5

[33] Tseng H-H, Luo Y, Ten Haken R K and El Naqa I 2018 The role of machine learning in knowledge-based response-adapted radiotherapy *Front. Oncol.* **8** 266

[34] Yang J, Sharp G C and Gooding M J 2021 *Auto-Segmentation for Radiation Oncology: State of the Art* (Boca Raton, FL: CRC Press)

[35] 2019 Advances in auto-segmentation *Seminars in Radiation Oncology* ed C E Cardenas, J Yang, B M Anderson, L E Court and K B Brock (Amsterdam: Elsevier)

[36] Yu C, Anakwenze C P, Zhao Y, Martin R M, Ludmir E B and S.Niedzielski J *et al* 2022 Multi-organ segmentation of abdominal structures from non-contrast and contrast enhanced CT images *Sci. Rep.* **12** 19093

[37] Baroudi H, Brock K K, Cao W, Chen X, Chung C and Court L E *et al* 2023 Automated contouring and planning in radiation therapy: what is clinically acceptable? *Diagnostics.* **13** 667

[38] Das I J, Moskvin V and Johnstone P A 2009 Analysis of treatment planning time among systems and planners for intensity-modulated radiation therapy *J. Am. Coll. Radiol.* **6** 514–7

[39] Nelms B E, Robinson G, Markham J, Velasco K, Boyd S and Narayan S *et al* 2012 Variation in external beam treatment plan quality: an inter-institutional study of planners and planning systems *Pract. Radiat. Oncol.* **2** 296–305

[40] Xiao Y, Cardenas C, Rhee D J, Netherton T, Zhang L and Nguyen C *et al* 2023 Customizable landmark-based field aperture design for automated whole-brain radiotherapy treatment planning *J. Appl. Clin. Med. Phys.* **24** e13839

[41] Huang K, Das P, Olanrewaju A M, Cardenas C, Fuentes D and Zhang L *et al* 2022 Automation of radiation treatment planning for rectal cancer *J. Appl. Clin. Med. Phys.* **23** e13712

[42] Purdie T G, Dinniwell R E, Letourneau D, Hill C and Sharpe M B 2011 Automated Planning of tangential breast intensity-modulated radiotherapy using heuristic optimization *Int. J. Radiat. Oncol. Biol. Phys.* **81** 575–83

[43] Kisling K, Zhang L, Shaitelman S F, Anderson D, Thebe T and Yang J *et al* 2019 Automated treatment planning of postmastectomy radiotherapy *Med. Phys.* **46** 3767–75

[44] Kisling K, Zhang L, Simonds H, Fakie N, Yang J and McCarroll R *et al* 2019 Fully automatic treatment planning for external-beam radiation therapy of locally advanced cervical cancer: a tool for low-resource clinics *J. Global Oncol.* **5** 1–9

[45] Hernandez S, Nguyen C, Parkes J, Burger H, Rhee D J and Netherton T *et al* 2023 Automating the treatment planning process for 3D-conformal pediatric craniospinal irradiation therapy *Pediatr. Blood Cancer* **70** e30164

[46] Appenzoller L M, Michalski J M, Thorstad W L, Mutic S and Moore K L 2012 Predicting dose–volume histograms for organs-at-risk in IMRT planning *Med. Phys.* **39** 7446–61

[47] Cornell M, Kaderka R, Hild S J, Ray X J, Murphy J D and Atwood T F *et al* 2020 Noninferiority study of automated knowledge-based planning versus human-driven optimization across multiple disease sites *Int. J. Radiat. Oncol. Biol. Phys.* **106** 430–9

[48] Rhee D J, Jhingran A, Huang K, Netherton T J, Fakie N and White I *et al* 2022 Clinical acceptability of fully automated external beam radiotherapy for cervical cancer with three different beam delivery techniques *Med. Phys.* **49** 5742–51

[49] Olanrewaju A, Court L E, Zhang L, Naidoo K, Burger H and Dalvie S *et al* 2021 Clinical acceptability of automated radiation treatment planning for head and neck cancer using the radiation planning assistant *Pract. Radiat. Oncol.* **11** 177–84

[50] Ford E, Conroy L, Dong L, de Los Santos L F, Greener A and Gwe-Ya Kim G *et al* 2020 Strategies for effective physics plan and chart review in radiation therapy: report of AAPM Task Group 275 *Med. Phys.* **47** e236–72

[51] Talcott W J, Lincoln H, Kelly J R, Tressel L, Wilson L D and Decker R H *et al* 2020 A blinded, prospective study of error detection during physician chart rounds in radiation oncology *Pract. Radiat. Oncol.* **10** 312–20

[52] Kisling K, Johnson J L, Simonds H, Zhang L, Jhingran A and Beadle B M *et al* 2019 A risk assessment of automated treatment planning and recommendations for clinical deployment *Med. Phys.* **46** 2567–74

[53] Nealon K A, Balter P A, Douglas R J, Fullen D K, Nitsch P L and Olanrewaju A M *et al* 2022 Using failure mode and effects analysis to evaluate risk in the clinical adoption of automated contouring and treatment planning tools *Pract. Radiat. Oncol.* **12** e344–53

[54] Nealon K A, Court L E, Douglas R J, Zhang L and Han E Y 2022 Development and validation of a checklist for use with automatically generated radiotherapy plans *J. Appl. Clin. Med. Phys.* **23** e13694

[55] Gronberg M P, Beadle B M, Garden A S, Skinner H, Gay S and Netherton T *et al* 2023 Deep learning-based dose prediction for automated, individualized quality assurance of head and neck radiation therapy plans *Pract. Radiat. Oncol.* **13** e282–91

[56] Puyati W, Khawne A, Barnes M, Zwan B, Greer P and Fuangrod T 2020 Predictive quality assurance of a linear accelerator based on the machine performance check application using statistical process control and ARIMA forecast modeling *J. Appl. Clin. Med. Phys.* **21** 73–82

[57] Ma M, Liu C, Wei R, Liang B and Dai J 2022 Predicting machine's performance record using the stacked long short-term memory (LSTM) neural networks *J. Appl. Clin. Med. Phys.* **23** e13558
[58] Chan M F, Witztum A and Valdes G 2020 Integration of AI and machine learning in radiotherapy QA *Front. Artif. Intell.* **3** 577620
[59] Zhao W, Shen L, Islam M T, Qin W, Zhang Z and Liang X *et al* 2021 Artificial intelligence in image-guided radiotherapy: a review of treatment target localization *Quant. Imag. Med. Surg.* **11** 4881–94
[60] Sher D J, Radpour S, Shah J L, Pham N-L, Jiang S and Vo D *et al* 2022 Pilot study of a wearable activity monitor during head and neck radiotherapy to predict clinical outcomes *JCO Clin. Cancer Inform.* **6** e2100179
[61] Rebelo N, Sanders L, Li K and Chow J C L 2022 Learning the treatment process in radiotherapy using an artificial intelligence-assisted chatbot: development study *JMIR Form. Res.* **6** e39443
[62] Hopkins A M, Logan J M, Kichenadasse G and Sorich M J 2023 Artificial intelligence chatbots will revolutionize how cancer patients access information: ChatGPT represents a paradigm-shift *JNCI Cancer Spectr.* **7** pkad010
[63] Poroch D 1995 The effect of preparatory patient education on the anxiety and satisfaction of cancer patients receiving radiation therapy *Cancer Nurs.* **18** 206–14
[64] Jimenez Y A, Cumming S, Wang W, Stuart K, Thwaites D I and Lewis S J 2018 Patient education using virtual reality increases knowledge and positive experience for breast cancer patients undergoing radiation therapy *Support. Care Cancer* **26** 2879–88
[65] McIntosh C, Conroy L, Tjong M C, Craig T, Bayley A and Catton C *et al* 2021 Clinical integration of machine learning for curative-intent radiation treatment of patients with prostate cancer *Nat. Med.* **27** 999–1005
[66] Isensee F, Jaeger P F, Kohl S A A, Petersen J and Maier-Hein K H 2021 nnU-Net: a self-configuring method for deep learning-based biomedical image segmentation *Nat. Methods* **18** 203–11
[67] Fedorov A, Beichel R, Kalpathy-Cramer J, Finet J, Fillion-Robin J-C and Pujol S *et al* 2012 3D Slicer as an image computing platform for the quantitative imaging network *Magn. Reson. Imaging* **30** 1323–41
[68] Korreman S, Eriksen J G and Grau C 2021 The changing role of radiation oncology professionals in a world of AI—just jobs lost—or a solution to the under-provision of radiotherapy? *Clin. Transl. Radiat. Oncol* **26** 104–7
[69] ACGME 2020 *Radiation Oncology Case Logs* https://apps.acgme-i.org/ads/Public/Reports/CaselogNationalReportDownload?specialtyId=98&academicYearId=26 (Accessed 4 March 2024)
[70] Sorin V, Barash Y, Konen E and Klang E 2020 Creating artificial images for radiology applications using generative adversarial networks (GANs)—a systematic review *Acad. Radiol.* **27** 1175–85
[71] McGinnis G J, Ning M S, Beadle B M, Joubert N, Shaw W and Trauernich C *et al* 2022 Barriers and facilitators of implementing automated radiotherapy planning: a multisite survey of low- and middle-income country radiation oncology providers *JCO Global Oncol.* **8** e2100431
[72] Union I T 2021 *Connectivity in the Least Developed Countries: Status Report 2021*

[73] Ajuwon G A 2015 Internet accessibility and use of online health information resources by doctors in training healthcare institutions in Nigeria *Libr. Philos. Pract* **2015** 1

[74] Grover S, Xu M J, Yeager A, Rosman L, Groen R S and Chackungal S *et al* 2015 A systematic review of radiotherapy capacity in low- and middle-income countries *Front. Oncol.* **4** 380

[75] Ngusie H S, Kassie S Y, Chereka A A and Enyew E B 2022 Healthcare providers' readiness for electronic health record adoption: a cross-sectional study during pre-implementation phase *BMC Health Serv. Res.* **22** 282

[76] Boshnak H, Gaber S A, Abdo A and Yehia E 2019 Guidelines to overcome the electronic health records barriers in developing countries *Int. J. Comput. Appl.* **975** 8887

Chapter 4

Low-cost, high-quality x-ray imaging technology for radiotherapy

Marios Myronakis and Ross Berbeco

4.1 Introduction

Image guidance in radiation therapy can save lives and reduce patient side effects [1]. Effective and safe radiation therapy depends on the accurate delivery of complex radiation fields to cancerous tumors. To provide the highest-quality radiation treatments, accurate knowledge of patient internal anatomy including the electron density makeup is essential. Image guidance for radiation therapy may not be consistently used or readily available in many cancer centers. Obstacles include upfront capital costs, ongoing personnel and maintenance costs, education and training, and patient throughput concerns. Even an initial sizable investment in state-of-the-art equipment could be insufficient if the other aspects are not addressed. An image-guided program that is not adequately supported may provide lower-quality treatments as well as risk more facility downtime while waiting for service, for example.

In this chapter, we describe an alternative to kilovoltage (kV) x-ray image guidance in radiation therapy. We have performed substantial research demonstrating that the megavoltage (MV) treatment beam itself may be used to perform the imaging procedures most often associated with kV imaging, including radiotherapy planning imaging, setup planar and volumetric imaging, adaptive planning, and in-treatment imaging. Here, we will focus on the volumetric imaging aspects, or MV cone-beam computed tomography (MV-CBCT).

4.2 Image-guidance technologies

There are multiple technological approaches to measuring patient position for image guidance each with various advantages and disadvantages (see table 4.1). The patient surface can be observed through surface-imaging approaches with several commercial products available. While these techniques have the advantage of

doi:10.1088/978-0-7503-3751-9ch4

Table 4.1. Comparison of different patient setup imaging technologies.

Technology	Radiation	Cost	Geometric accuracy	HU accuracy
Optical/surface	None	Low	Low	N/A
MRI	None	Highest	High	Low
PET	Low	High	Medium	None
kV x-rays	Low	Medium	High	Medium
MV x-rays	Low	Low	High	High

simplicity, safety, and low cost, there is a significant disadvantage of trusting the external anatomy to reliably report the status and whereabouts of the internal anatomy [2]. Magnetic resonance imaging (MRI) has recently been introduced for radiotherapy image guidance [3]. This modality offers the best anatomy visualization, including real-time tracking, but is hugely expensive and has no facility for accurate electron density estimation. The marriage of positron emission tomography (PET) with a medical linear accelerator also has a high cost, including the ancillary facilities needed to manage the radioisotopes, and is not yet proven as a benefit for online imaging. The use of on-board kV x-ray imaging has become standard of care and offers a compromise in terms of cost and clinical benefit.

4.3 MV imaging—challenges and opportunities

MV x-ray imaging stands out among the options above for anatomical imaging without substantial technology or infrastructure investment. The physics of MV imaging have inherent challenges and opportunities, both of which can be addressed through technical innovation.

MV imaging uses photons generated by the treatment linear accelerator (figure 4.1). Therefore, the treatment beam itself can be used for imaging. Since this beam was not designed for imaging, small modifications of the beam energy can improve imaging without the need for additional equipment. Innovations in flat-panel detector technology can improve image quality, reducing the soft-tissue contrast disadvantages relative to kV imaging.

The inherent physics of MV photon interactions in the patient and in the detector are less favorable than for kV photons. Poor efficiency of MV imaging leads to low photon counts in the detector resulting in noisy images. The interactions of higher-energy photons within the patient anatomy are less dependent on the atomic number (Z) of the anatomy than lower-energy photons due to fewer photoelectric interactions. This leads to less contrast, particularly for bony anatomy and implanted radiopaque fiducials.

A special advantage of MV imaging is the opportunity to visualize the treated anatomy as it is being treated. This 'beam's-eye-view' approach provides real-time two dimensional (2D) images, showing exactly which anatomy is being irradiated at

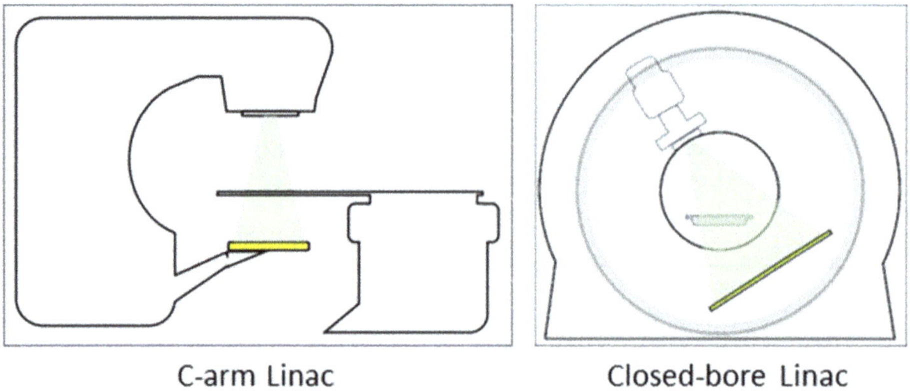

Figure 4.1. MV imagers are located in the MV treatment beam line to collect the exit radiation that has passed through the patient. Illustrations are shown for (left) an open C-arm type linear accelerator system and (right) a closed-bore linear accelerator.

every moment [4]. The images may be reviewed offline or analyzed online with the information used for real-time tumor tracking [4–6].

4.4 MV-CBCT

Most modern radiotherapy treatment units have a flat-panel detector mounted for portal imaging. That is a planar 2D projection used for patient setup and positioning. Modern treatment applications such as intensity-modulated radiotherapy and volumetric modulated arc therapy require increased accuracy in patient setup and positioning to adhere to strict tumor and organ-at-risk treatment plan conformality.

The primary aim of MV cone beam–computed tomography (CBCT) is to assist in the accurate and fast positioning and setup of the patient for radiotherapy treatment. MV-CBCT acquisitions use the radiotherapy beam (MV energies) and a flat-panel detector mounted on the gantry of the treatment machine. During MV-CBCT acquisition, the patient is placed on the treatment couch and multiple projections are acquired with open field over an arc range between 200 and 360 degrees [7]. After the acquisition, a three-dimensional volume of the patient is generated and registered with the treatment planning CT to determine setup shifts.

The 6 MV treatment beam is commonly used for MV-CBCT acquisitions. Higher energies will have a negative impact on subject contrast and are generally not used for imaging. Recent implementations in commercial treatment units utilize 2.5 MV beams specifically for imaging [8, 9]. The 2.5 MV beams contain a higher proportion of photons with energies below 80 keV that inherently enhance subject contrast and are easier to detect with conventional flat-panel detectors due to increased probability of photoelectric absorption in tissues and the detector material.

The flat-panel detectors used in MV-CBCT imaging are constructed in part from rare-earth materials such as gadolinium and terbium. The most common compound found in commercial flat-panel detectors mounted on the treatment gantry is terbium dopped gadolinium oxysulfide (Gd_2O_2S:Tb or GOS). GOS flat-panel detectors are relatively inexpensive but have low detection capabilities in the MV energy range leading to deteriorated performance in terms of image noise, soft-tissue contrast, and required dose to achieve a certain image quality [10]. The panel is placed at 150–160 cm distance from the treatment head as this was found as the optimal range to reduce scatter contamination in the images [11].

In treatment units with a cylindrical (ring-like, closed-bore) gantry (e.g. TomoTherapy®, Accuray) an alternative to MV-CBCT is pretreatment CT with the MV beam [12]. The energy of the beam is usually 6 MV without flattening filter. The number of x-rays with low energy are not filtered and subsequently enhance subject contrast compared to the filtered 6 MV beams used in C-arm gantry types. The flat panel is positioned at 150 cm from the treatment head. Early implementations were using thick thallium doped cesium iodide (CsI:Tl) detectors to increase detection efficiency [13].

Novel detector designs, with different detector geometry (e.g. increased thickness) and/or different low-cost materials are under research to improve noise properties and soft-tissue contrast [14–17].

The applications of MV-CBCT are not limited to patient setup and positioning. The feasibility of MV-CBCT images for urgent treatment planning has been evaluated with success, albeit challenges with field of view, patient size, and peripheral tumor locations remain [18, 19].

4.5 Novel developments

4.5.1 Multilayer imager design

Detector efficiency for MV imaging can be improved by increasing the effective scintillator thickness (or interaction cross section) for greater photon collection. Strategies include: (1) stacking of identical layers, (2) using novel, low-cost, high-efficiency scintillating materials, and (3) stacking combinations of low-efficiency, high-resolution scintillator layers with high efficiency, low-resolution layers.

A multilayer imager (MLI) has been built combining four identical imager layers, each one equivalent to a standard single-layer imager [15]. MV-CBCT data was acquired with the MLI at radiation dose levels equivalent to kV-CBCT protocols [20]. Figure 4.2 shows a comparison between CBCT reconstructions acquired with the standard clinical kV on-board imaging (OBI) system, the standard clinical AS1200 MV imager, and the combined four layers of the MLI. The radiation dose levels were nominally equivalent across each imaging modality. The MLI significantly decreases noise, increasing image quality and detectability of low-contrast materials, compared to MV-CBCT with the standard single-layer detector. It was further found that MV-CBCT with the MLI provided much improved HU accuracy compared to kV-CBCT, indicating that it is better suited for radiotherapy dose planning. This has important implications for cancer centers with limited resources or inconsistent access to CT simulators.

The impact of scintillator composition and thickness on MV-CBCT quality has been studied in analytical modeling studies [21, 22] and experimental measurements [14]. In general, increased photon detection efficiency leads to decreases in noise for

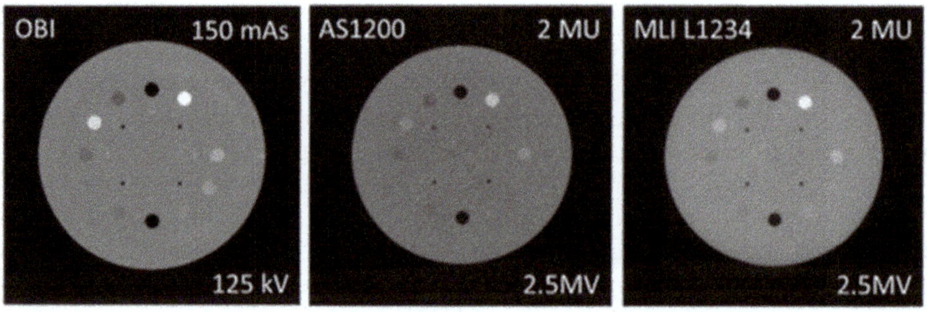

Figure 4.2. Improvements in image quality are shown for the four-layer MLI. Reconstructed material insert slices are shown for (left) the standard kV OBI and 125 kVp source, (middle) the standard single-layer MV detector (AS1200), and (right) the four-layer MLI with all layers combined and 2.5 MV beam delivery. All images were acquired at nominally the same dose (Myronakis M *et al* Med. Phys. 2020). Adapted from [20] John Wiley & Sons.

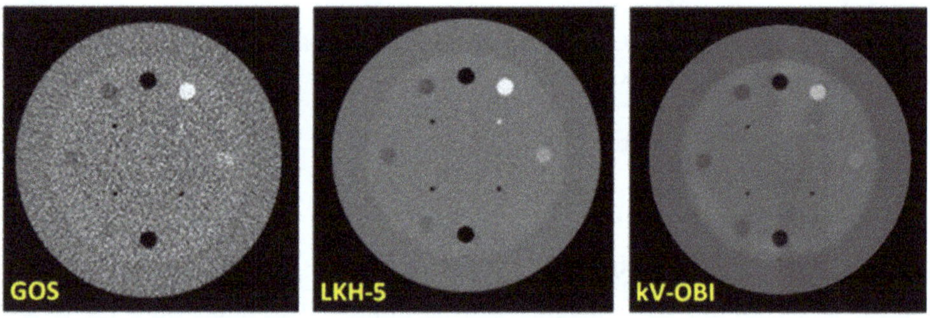

Figure 4.3. The improved noise performance of a novel scintillating glass (LKH-5) enables MV-CBCT image quality approaching that of kV-CBCT, at the same imaging dose (2.7 cGy). (left) The low detection efficiency of a standard commercial single-layer MV detector employing a thin GOS scintillator leads to high-noise MV-CBCT acquisitions, obscuring low-contrast features. (middle) Increased detector efficiency leads to a major reduction in noise relative to the standard GOS MV detector, revealing the low-contrast features. (right) kV-CBCT acquisition with a commercial OBI system is shown for comparison at the same imaging dose level. Reprinted from [14] John Wiley & Sons.

improved contrast and feature detection. Experiments with a novel scintillating glass (LKH-5) demonstrated MV-CBCT image quality approaching kV-CBCT, for the same imaging dose (figure 4.3). It was also found that the reduced spatial resolution from thicker scintillators has a negligible effect on feature detection for radiotherapy applications. The improvements in noise performance far outweigh the reduction in resolution, providing a clear direction for further advances.

4.5.2 Electron density mapping for Adaptive Radiation Therapy (ART)

Apart from setup and positioning, MV-CBCT can have more involvement in patient treatment. One of the major advantages of MV-CBCT is that the patient is on the treatment couch and the images are acquired at the same (or similar) energy with the treatment beam. This is attractive for situations where adaptive treatments are considered. Without moving the patient to a different modality, on-site MV-CBCT reconstructions may be used for treatment planning. Major challenges that need to be addressed to achieve such tasks are corrections of beam hardening and scatter in the projections of an MV-CBCT acquisition. Early studies demonstrated the feasibility to calibrate MV-CBCT for accurate estimation of electron densities needed for treatment planning [23–25].

4.6 Volumetric imaging dose

Although the dose delivered to the patient from imaging procedures is a small fraction of the received treatment dose it should not be neglected altogether. Common treatment schemes include 25–40 fractions and last several weeks. The potential cumulative dose from repeated MV-CBCT acquisitions during inter- and intrafraction adaptive treatments over the whole treatment course may substantially increase deterministic effects, and the risk for late stochastic effects, especially in younger patients.

MV-CBCT dose with conventional flat-panel imagers is in the order of 50–150 mGy [26, 27] depending on anatomical site. For reference, the established threshold for the appearance of deterministic effects from radiation exposure is 100 mGy [28]. Lowering MV-CBCT dose is an active area of research. Recent developments using novel flat-panel designs combined with a 2.5 MV beam provided an average threefold increase in contrast-to-noise ratio compared with a conventional flat panel at nearly 20 mGy delivered dose (figure 4.2) [20].

4.7 Cost savings with MV imaging

While the applications described above could be beneficial in any environment, there are specific attributes of MV-CBCT that make it especially attractive for low-resource settings. These include the minimization of the financial burden of the initial equipment purchase and ongoing maintenance, and expanded functionality as a radiotherapy simulator, if needed.

While it is difficult to put a dollar amount on the initial financial outlay for kV-CBCT, the rough cost of the physical components (e.g. x-ray tube, generator, and panel) is likely several hundreds of thousands of dollars. This calculation does not include the computer hardware or software used for processing and displaying the images as these will also be needed for MV imaging. For our calculations, these same pieces are still needed. Regarding ongoing maintenance, beyond the financial expense there is also the possibility of extended downtime while waiting for parts or service support, especially in remote locations.

For clinics in which a CT is not readily available for treatment planning simulation, due to scheduling, logistical, or other challenges, on-board MV-CBCT could be used instead. As explained above, the associated electron density maps should be well suited for dose calculation. Relative to fan–beam kV-CT imaging, MV-CBCT will also have improved metal–artifact reduction capabilities [29] which will be important for lesion delineation in many circumstances. In other cases,

MV-CBCT will be inferior to kV-CT. However, the differences can be mitigated by novel imager technologies, described above, to reduce noise and employing an MV fan beam to reduce scatter [30].

4.8 Conclusion

Low-cost, high-quality, reliable imaging can be made available through minor changes to existing technology. With access to this technology, clinics in low-resource environments can deliver high-accuracy image-guided radiotherapy, on par with clinics that have more access to resources. Additional benefits could include reduced downtime and other costs associated with auxiliary kV x-ray imaging systems. In addition, this technology does not shut clinics out of emerging novel treatment approaches, like adaptive radiotherapy. The patients will benefit from the high-quality imaging and reduced downtime. Overall, MV imaging represents an important opportunity for low-cost, high-quality radiotherapy imaging.

Acknowledgments

Research reported in this chapter was supported, in part, by grants from Varian Medical Systems, Inc. and the National Cancer Institute of the National Institutes of Health under award number R01CA188446. The content is solely the responsibility of the authors and does not necessarily represent the official views of the National Institutes of Health.

References

[1] Bujold A, Craig T, Jaffray D and Dawson L A 2012 Image-guided radiotherapy: has it influenced patient outcomes? *Semin. Radiat. Oncol.* **22** 50–61

[2] Ionascu D, Jiang S B, Nishioka S, Shirato H and Berbeco R I 2007 Internal-external correlation investigations of respiratory induced motion of lung tumors *Med. Phys.* **34** 3893–903

[3] Goodburn R J *et al* 2022 The future of MRI in radiation therapy: challenges and opportunities for the MR community *Magn. Reson. Med.* **88** 2592–608

[4] Berbeco R, Hacker F, Ionascu D and Mamon H 2007 Clinical feasibility of using an EPID in cine mode for image-guided verification of SBRT *Int. J. Radiat. Oncol. Biol. Phys.* **69** 258–66

[5] Rottmann J, Keall P J and Berbeco R I 2013 Makerless EPID image guided dynamic multileaf collimator tracking for lung tumors *Phys. Med. Biol.* **58** 4195–204

[6] Rottmann J, Keall P J and Berbeco R I 2013 Real-time soft tissue motion estimation for lung tumors during radiotherapy delivery *Med. Phys.* **40** 091713

[7] Morin O, Gillis A, Chen J, Aubin M, Bucci M K, Roach M and Pouliot J 2006 Megavoltage cone-beam CT: System description and clinical applications *Med. Dosim.* **31** 51–61

[8] Gräfe J L, Owen J, Eduardo Villarreal-Barajas J and Khan R F 2016 Characterization of a 2.5 MV inline portal imaging beam *J. Appl. Clin. Med. Phys.* **17** 222–34

[9] Ding G X and Munro P 2017 Characteristics of 2.5 MV beam and imaging dose to patients *Radiother. Oncol.* **125** 541–7

[10] Antonuk L E 2002 Electronic portal imaging devices: a review and historical perspective of contemporary technologies and research *Phys. Med. Biol.* **47** R31–65

[11] Jaffray D A, Siewardsen J H, Wong J W and Martinez A A 2002 Flat-panel cone-beam computed tomography for image-guided radiation therapy *Int. J. Radiat. Oncol. Biol. Phys.* **53** 1337–49

[12] Ruchala K J, Olivera G H, Schloesser E A and Mackie T R 1999 Megavoltage CT on a tomotherapy system *Phys. Med. Biol.* **44** 2597–621

[13] Mosleh-Shirazi M A, Evans P M, Swindell W, Webb S and Partridge M 1998 A cone-beam megavoltage CT scanner for treatment verification in conformal radiotherapy *Radiother. Oncol.* **48** 319–28

[14] Hu Y H *et al* 2019 Characterizing a novel scintillating glass for application to megavoltage cone-beam computed tomography *Med. Phys.* **46** 1323–30

[15] Rottmann J, Morf D, Fueglistaller R, Zentai G, Star-Lack J and Berbeco R 2016 A novel EPID design for enhanced contrast and detective quantum efficiency *Phys. Med. Biol.* **61** 6297–306

[16] Harris T C *et al* 2020 Clinical translation of a new flat-panel detector for beam's-eye-view imaging *Phys. Med. Biol.* **65** 225004

[17] Seppi E J *et al* 2003 Megavoltage cone-beam computed tomography using a high-efficiency image receptor *Int. J. Radiat. Oncol. Biol. Phys.* **55** 793–803

[18] Thomas T H, Devakumar D, Purnima S and Ravindran B P 2009 The adaptation of megavoltage cone beam CT for use in standard radiotherapy treatment planning *Phys. Med. Biol.* **54** 2067–77

[19] Held M, Sneed P K, Fogh S E, Pouliot J and Morin O 2015 Feasibility of MV CBCT-based treatment planning for urgent radiation therapy: dosimetric accuracy of MV CBCT-based dose calculations *J. Appl. Clin. Med. Phys.* **16** 458–71

[20] Myronakis M *et al* 2020 Low-dose megavoltage cone-beam computed tomography using a novel multi-layer imager (MLI) *Med. Phys.* **47** 1827–35

[21] Hu Y H *et al* 2018 Physics considerations in MV-CBCT multi-layer imager design *Phys. Med. Biol.* **63** 125016

[22] Hu Y H *et al* 2018 Leveraging multi-layer imager detector design to improve low-dose performance for megavoltage cone-beam computed tomography *Phys. Med. Biol.* **63** 035022

[23] Petit S F, van Elmpt W J, Nijsten S M, Lambin P and Dekker A L 2008 Calibration of megavoltage cone-beam CT for radiotherapy dose calculations: correction of cupping artifacts and conversion of CT numbers to electron density *Med. Phys.* **35** 849–65

[24] Hughes J, Holloway L C, Quinn A and Fielding A 2012 An investigation into factors affecting electron density calibration for a megavoltage cone-beam CT system *J. Appl. Clin. Med. Phys.* **13** 3271

[25] Morin O, Chen J, Aubin M, Bose S, Gillis A, Bocci M and Pouliot J 2005 Dose calculation using megavoltage cone beam CT imaging *Int. J. Radiat. Oncol. Biol. Phys.* **63** 103

[26] Pouliot J *et al* 2005 Low-dose megavoltage cone-beam CT for radiation therapy *Int. J. Radiat. Oncol. Biol. Phys.* **61** 552–60

[27] Morin O, Gillis A, Descovich M, Chen J, Aubin M, Aubry J F, Chen H, Gottschalk A R, Xia P and Pouliot J 2007 Patient dose considerations for routine megavoltage cone-beam CT imaging *Med. Phys.* **34** 1819–27

[28] The 2007 Recommendations of the International Commission on Radiological Protection. ICRP publication 103 *Ann. ICRP* 2007 **37** 1–332
[29] Harris T C *et al* 2023 Impact of a novel multilayer imager on metal artifacts in MV-CBCT *Phys. Med. Biol.* **68** 145009
[30] Gong H, Tao S, Gagneur J D, Liu W, Shen J, McCollough C H, Hu Y and Leng S 2021 Implementation and experimental evaluation of Mega-voltage fan-beam CT using a linear accelerator *Radiat. Oncol.* **16** 139

IOP Publishing

Humanitarian Engineering for Global Oncology

Eric Ford

Chapter 5

Engineering smart biomaterials for hypofractionated radiotherapy

Lensa Keno, Michele Moreau, Sayeda Yasmin-Karim and Wilfred Ngwa

5.1 Introduction

One of the main ways to treat cancer is radiotherapy, which uses radiation to kill cancer cells. It is used for more than half of the people with cancer. Sometimes, doctors put inactive materials inside the tumors to help them aim the radiation better. These are called radiation therapy (RT) biomaterials and they only have one job: to make sure the radiation hits the right spot. But what if these materials could do more than that? Some researchers have suggested that we could make 'smart' RT biomaterials that could react to something and do extra things like release drugs or other substances that could make the radiation work better and cause less harm to healthy tissues. There are also other kinds of smart RT biomaterials, such as nanoparticles that can be turned on to enhance radiotherapy. This chapter covers the engineering of smart RT biomaterials, the challenges and opportunities for further research, and some of the possible ways to use them in the clinic. Some of these ways include making cancer treatment more effective and less harmful, combining radiotherapy with other treatments like immunotherapy or chemotherapy, saving time or money in treatment, and other new ways to make them accessible in low-resource settings.

Cancer can be treated in different ways, such as radiotherapy, surgery, chemotherapy, and immunotherapy. RT [1] is a way of using radiation to kill cancer cells and it is used for more than half of the people with cancer. However, there are still some problems with these treatments. For example, radiation can damage healthy tissues and cause side effects like nerve pain, memory loss, kidney problems, hearing loss, and more. To avoid this, doctors use some materials that they put inside the tumors to help them aim the radiation better and protect the healthy tissues. These materials are called radiotherapy biomaterials and they do not do anything else. But what if they could do more?

doi:10.1088/978-0-7503-3751-9ch5

Some researchers have suggested that we could engineer smart radiotherapy biomaterials (SRBs) or biomaterial drones that could do more than one thing [2, 3]. For example, they could help with aiming the radiation, make the radiation work better, boost the immune system, or deliver drugs to the tumor. These SRBs could be engineered by changing some of the materials that are already used for radiotherapy, such as fiducials or beacons. These changes could be made by using some substances like poly(lactic-co-glycolic acid) (PLGA), chitosan, or carbohydrate polymers and high atomic number nanoparticles. These SRBs could deliver immunotherapy to the tumor area and avoid affecting the rest of the body. This way, they could create a vaccine inside the tumor that could teach the immune system to recognize and attack cancer cells. This is called *in situ* vaccination. This would be better than traditional cancer vaccines that take a long time to make and are very expensive.

SRB-mediated *in situ* vaccination would be faster and cheaper. *In situ* vaccination with radiotherapy can also cause something called the abscopal effect, which means that radiotherapy of one tumor can make other tumors shrink, even if they are not treated with radiation. This is especially useful for tumors that have spread to other parts of the body, which cause most cancer deaths. In this chapter, we will cover how radiotherapy biomaterials have been engineered for hypofractionated radiotherapy (HFRT), e.g. for prostate, lung, pancreatic, cervical, and other cancers, and how SRBs can be used for radio-immunotherapy that combines radiotherapy and immunotherapy. We will also discuss clinical translation of such SRBs and other possibilities for research and development, such as new ways of doing radiotherapy, such as FLASH radiotherapy.

5.2 SRBs

SRBs [3] can do more than one thing, like helping with aiming the radiation, making the radiation work better, changing the tumor environment, and delivering drugs or other substances to the tumor. These drugs or substances can include chemotherapy, immunotherapy, or nanoparticles that can make the tumor cells more sensitive to radiation. Some of these nanoparticles are made of elements that have a high atomic number (high-Z), which means they can show up better on imaging and also produce more electrons when they are hit by radiation. These electrons can damage the cancer cells more within a short distance. Some examples are gold particles (GNPs) and gadolinium particles (GdNPs), which can be seen on both magnetic resonance imaging (MRI) and computed tomography (CT) scans. Figure 5.1 highlights the different types of SRBs currently in preclinical research on clinical translation with potential to enhance HFRT, and even reduce treatment costs substantially when combining with immunotherapy.

The first types of SRBs under development are seed-like, and are similar to solid fiducials. Figure 5.1(A) illustrates the design of such SRBs to provide image contrast and radiation enhancement as well as delivery of drugs. The seed SRB can be created with a mix of PLGA polymer and formed to the size and shape of current fiducials. As can be seen in figure 5.1(A), high-Z nanoparticles can either be loaded together with drugs and incorporated in the hollow core of the SRB or the

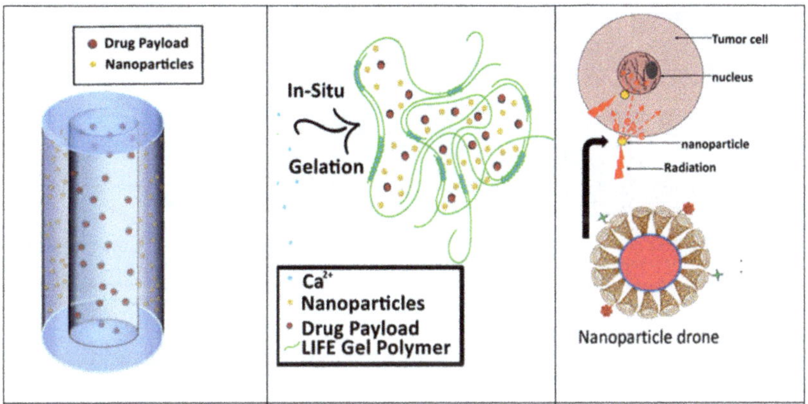

A: Design of Seed-like smart radiotherapy biomaterial (SRB) the size of a grain of rice, which can be administered as fiducials currently used in clinic.	B: Design of gel-like smart radiotherapy biomaterial (SRB), which can be administered as fiducials currently used in clinic.	C: Design of smart radiotherapy biomaterial (SRB in the form of colloidal nanoparticle drones.

Figure 5.1. Types of SRBs currently in preclinical research on clinical translation with potential to enhance HFRT.

nanoparticles can be incorporated within the biodegradable polymer matrix of the SRB.

The second type of SRB under development is in the form of a liquid fiducial. The liquid SRB technology is made of solution of sodium alginate and chitosan that is rapidly transformed into an anchored hydrogel in the presence of calcium ions (Ca^{2+}) within the tumor (figure 5.1(B)). Immunotherapy drug payloads or high-Z nanoparticles can be incorporated into this biodegradable polymer to enhance image contrast. This enables a potent amount of immunotherapeutic agents to diffuse locally for a sustained period within the tumor with minimal leakage into neighboring organs. This represents a potential new generation of multifunctional liquid fiducials, combining image guidance with *in situ* drug delivery, to engender effective *in situ* vaccination.

A third type of SRB that has been engineered for HFRT applications combining with immunotherapy is the nanoparticle drone or nanodrone. Nanodrones loaded with immunotherapy drugs and with a high-Z component can be administered to accumulate in the tumor. Examples include nanodrones made of gadolinium or gold or a hybrid, which permit the tracking of this accumulation via CT or MRI. Such platforms can be optimized for image-guided drug delivery, which could allow for quantification of distribution in tumors over time. In general, nanodrones that carry immunotherapy to boost the immune system have many advantages. They can protect the drug from degradation and toxicities and make it more potent, as well as deliver it to specific cells and release it slowly. However, more research is ongoing on this. Using nanodrones with radiotherapy can trigger the abscopal effect during radiotherapy.

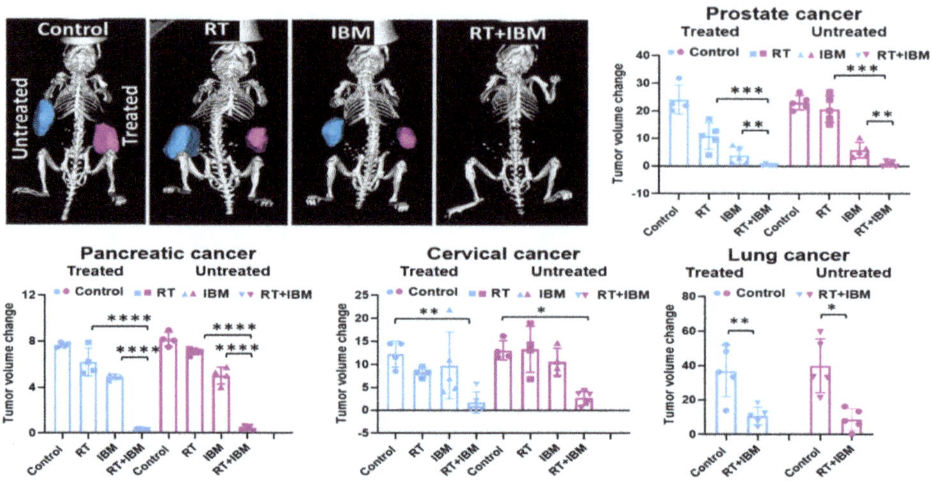

Figure 5.2. Results in animal studies illustrating the potential of SRBs loaded with immunoadjuvants (IBMs) loaded with immunoadjuvants for boosting the abscopal effect or *in situ* vaccination causing regression of both treated and distant untreated tumors (e.g. metastasis). Reproduced from [5]. CC BY 4.0.

Figure 5.2 highlights some results when using SRBs to boost the abscopal effect (see [4, 5] and references for more details). The results are shown in animal studies for different cancers, highlighting the potential for such an *in situ* vaccination approach working across different tumor types.

Further investigations are needed to optimize the use of SRBs for clinical trial. Already the FDA has approved a clinical trial, in an approach which could present opportunity for not only smart biomaterials to enhance HFRT, but to combine this with immunoadjuvants to make low-cost immunotherapy possible. With the high level of localization provided by the SRBs, more concentrated doses of loaded therapy can be delivered directly to the tumor, avoiding the toxicities related to systemic distribution that have hampered clinical translation of some immunotherapies.

5.3 Potential benefits in global health

Pervasive disparities exist in access to RT, especially in low- and middle-income countries (LMICs). The recent World Health Organization Cancer Report describes the growing global burden of cancer and disparities as alarming, with over 70% of 18.1 million cases and 60% of 9.6 million deaths per year occurring in LMICs. In African LMICs, where survival rates are amongst the lowest, the cancer burden is projected to reach up to 1 million deaths annually by 2030 per a new Lancet Oncology Commission chaired by the principal investigator. The primary drivers of these disparities in incidence and deaths may underlie the higher cancer-related morbidity and mortality rates also seen among African American/immigrant populations in the United States of America. Innovative approaches to reduce these disparities in access to RT are crucial in addressing the growing global burden of cancer deaths and associated disparities.

Innovative approaches with potential to significantly reduce disparities in access to RT include use of evidence-based HFRT and SRBs. Major benefits of HFRT, which involves the delivery of larger doses of radiation per treatment fraction in order to complete the full course of treatment over a shorter period of time, include: (1) more patients may have access to treatment since each patient can come in fewer times for radiotherapy compared to conventional treatment, e.g. 7 times/fractions instead of 39 times for prostate cancer; (2) increased patient convenience; and (3) expected cost savings given the fewer number of fractions. Hence, the use of HFRT with SRBs has major potential for increasing access to RT.

Overall the engineering of SRBs has major potential to increase access to RT and establish the potential of new approaches like the use of HFRT and SRBs in combining radiotherapy and immunotherapy, and reducing the pervasive disparities in access to RT in Africa and amongst African Americans, driven by such factors as economics and treatment time.

References

[1] Atun R, Jaffray D A and Barton M B *et al* 2015 Expanding global access to radiotherapy *Lancet Oncol.* **16** 1153–86

[2] Yasmin-Karim S *et al* 2022 Boosting the abscopal effect using immunogenic biomaterials with varying radiation therapy field sizes *Int. J. Radiat. Oncol. Biol. Phys.* **112** 475–86

[3] Ngwa W, Irabor O C, Schoenfeld J D, Hesser J, Demaria S and Formenti S C 2018 Using immunotherapy to boost the abscopal effect *Nat. Rev. Cancer* **18** 313–22

[4] Moreau M, Yasmin-Karim S and Kunjachan S *et al* 2018 Priming the abscopal effect using multifunctional smart radiotherapy biomaterials loaded with immunoadjuvants *Front. Oncol.* **8** 56

[5] Yasmin-Karim S, Wood J, Wirtz J, Moreau M, Bih N, Swanson W, Muflam A, Ainsworth V, Ziberi B and Ngwa W 2021 Optimizing *in situ* vaccination during radiotherapy *Front. Oncol.* **11** 711078

IOP Publishing

Humanitarian Engineering for Global Oncology

Eric Ford

Chapter 6

Humanitarian engineering solutions: surgical oncology

Beatrice Wiafe Addai and Lawrencia Dsane Bawuah

Surgery is a very important treatment modality in cancer care, either as the primary treatment for cure, as downstaging of the tumor prior to other therapeutic options, or as a palliative care option. However, many individuals in resource-poor settings are unable to access surgical services. It is important to understand the barriers that impede surgical management in cancer care in resource-limited settings like in Africa. Humanitarian engineering solutions for surgical oncology can have a significant impact in overcoming these barriers and reducing the global burden of cancer, especially in low- and middle-income countries (LMICs) where access to quality cancer care is limited. Such solutions can be critical for women in particular in African LMICs, where breast and cervical cancers are leading cancers and surgical oncology is an important part of the treatment options. Here we discuss the barriers and potential humanitarian engineering solutions to address these barriers.

6.1 Barriers: the perceived barriers to surgical treatment in cancer management in LMICs, with focus on Africa

One commonly cited barrier to care in LMICs which we have identified is fear of surgery. Sometimes surgery is only considered as a viable option if all other alternatives have proven futile. Unfortunately, this causes delays and makes an operable condition inoperable, with the biggest reason being fear of worse outcomes, fear of pain, and fear of losing an organ, which might result in deformity, leading to divorce or separation. The misunderstandings surrounding anesthesia are also a great problem; some patients are afraid they might not survive surgeries or die during surgery due to wrong anesthesia.

One of the frequently asked questions posed to a surgeon is 'are you sure there is no other alternative to surgery?' In most African countries patients resort to prayer camps/spiritual healers, herbalists, or traditional healers instead of seeking medical

doi:10.1088/978-0-7503-3751-9ch6

help. This choice is mostly preceded by lack of knowledge and the fear of surgery. This suggests that cultural traditions also play a key role in the choice to refuse surgical treatment. The picture painted now shows that this barrier is a result of the lack of awareness that their cancers could be cured/treated through surgical intervention. As of 2020, the leading cancers reported in Africa were breast, cervix, prostate, liver, and colorectal cancers (data source: GLOBOCAN 2020). These five cancers account for almost 50 percent of all the reported cancer cases (figure 6.1).

All these cancers have surgery as a main component of the complete treatment. This makes improving surgical healthcare important.

In these societies women are normally the caregivers in the family, and often times ignore their own complaints; they normally assume the superman hero role and are more likely to feel like a burden to their families when they avail themselves of surgery. Some women have to ask permission from their husbands or the men in their family prior to getting any form of surgical treatment. Even though that might be the only curative method, some men in Africa can restrict their wives from undergoing mastectomy or hysterectomy. In order to improve this condition, women must be empowered to be able to decide for themselves when it most certainly involves their own health. Awareness creation about the cancers and the available treatment modalities should be carried out across the board, so men can support their wives or female relatives when there is a decision to be made about surgical treatment as a standalone treatment or complementary treatment in cancer care.

A serious human resource barrier is the shortage of well-trained professionals, i.e. the availability of adequately trained and centrally-placed professionals in the surgical supply chain system (doctors, nurses, anesthetists, psychologists, good

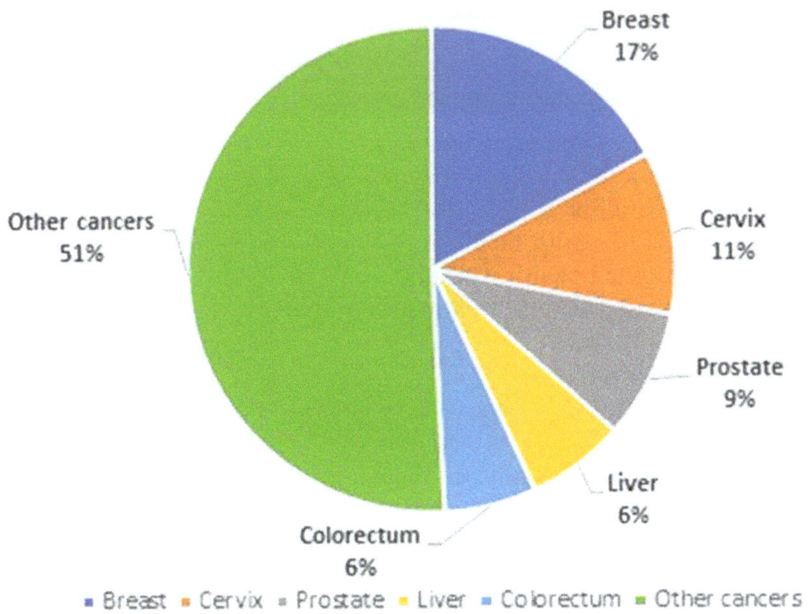

Figure 6.1. Estimated number of new cancer cases in Africa (male and female).

patient navigation systems, etc) in cancer management. According to this report by the Lancet Oncology [1] surgery must be at the heart of global and national cancer control planning.

The absence of social support systems in LMICs is a huge barrier. Almost all the treatment centers are based in the cities. This requires patients to sometimes travel for hours and even days to get to the facility where they can get the needed surgery. Unfriendly environments, lack of proper navigation systems for patients, and the lack of functioning social support systems make access to healthcare in some situations impossible. In the case of some women, not finding reliable persons or family members to take care of their children while they undergo treatment can be a huge barrier.

It is a well-known fact that the cost of cancer treatment is the primary barrier to receiving surgery. Other indirect costs such as travel expenses and accommodation in the area of the hospital are very prominent among worries and therefore cause a barrier to access surgical treatment. The direct costs of the surgery itself are for most of these patients not affordable and they are therefore dependent on their families for assistance when it comes to paying for the bills.

A way out would be to support patients who need surgical treatments financially. Many patients have to borrow from relatives in order to cover the costs, suggesting that the ability to pay for surgery is also contingent upon a relative's perception of the value of surgery. In rural agricultural settings, income is mostly seasonal and therefore money to pay for surgery is not always readily available for the family until their crops are sold, suggesting that time of year plays a critical role in a patient's ability to pay for cancer surgery. Overall, in resource-poor nations, both indirect and direct cost are major barriers to receiving surgical care in cancer treatment.

For a broader perspective, according to a systematic review of the literature [2], some of the common barriers to cancer care in LMICs that also apply to surgical oncology are:

- Low health literacy
- Lack of awareness and knowledge
- Stigma and fear
- Financial constraints
- Limited access to health facilities
- Shortage of trained health workers
- Inadequate diagnostic and treatment equipment
- Poor referral systems
- Long waiting times
- Cultural and religious beliefs.

These barriers can cause significant delays in the cancer care continuum, from accessing and diagnosing to treating cancer patients. The delays can vary depending on the type of cancer, the income level of the country, and the geographic region where the patient resides.

Another systemic review of the literature [3] elaborates the psychosocial influences on help-seeking behavior for cancer in low-income and lower middle-income countries. The use of traditional, complementary, and alternative medicine and gender influences appear to be important barriers to help-seeking in LMICs. This makes awareness creation a needed part in solving these challenges.

6.2 Potential humanitarian engineering solutions

Some possible interventions to overcome barriers to surgical oncology include the following:

- Increasing number of public education programs and awareness campaigns on the various cancers
- Providing affordable and accessible screening and treatment services
- Strengthening health system capacity and quality
- Improving referral and follow-up mechanisms
- Engaging community and religious leaders
- Addressing social determinants of health [4].

As an example, Breast Care International (BCI) a nongovernmental organization (NGO) that works to improve breast cancer awareness and care in Ghana has collaborated with Peace and Love Hospitals, which are specialized in breast cancer diagnosis and treatment. Some of the ways that BCI has promoted public education and awareness are:

- Conducting community outreach programs in deprived areas, where they educate the community about breast and cervical cancer. Manually screening women for signs of breast cancer and providing referrals and follow-ups.
- Organizing annual walks for the cure, where they mobilize thousands of people to join the campaign against breast cancer and raise funds for their activities.
- Training community breast health promoters and peer navigators, who are volunteers that educate their peers about breast cancer prevention, detection, and treatment, and provide emotional and practical support to patients.
- Training community health workers on cervical cancer and detection of cervical abnormalities.
- Publishing research articles and reports on the impact of their programs on the knowledge, attitudes, and practices of women towards breast cancer in Ghana.
- Engaging with media, policy makers, civil society, and other stakeholders to advocate for better breast and cervical cancer policies and better health services.

Humanitarian engineering is the application of engineering to improving the well-being of marginalized people and disadvantaged communities, usually in LMICs. Some of the ways that BCI and Peace and Love Hospitals have promoted public education and awareness are relevant to humanitarian engineering, such as:

- Providing affordable and accessible screening and treatment services, which involves applying engineering skills to design and implement appropriate technologies and infrastructure for healthcare delivery.
- Strengthening health system capacity and quality, which involves applying engineering skills to optimize the performance and efficiency of health facilities, equipment, and personnel.
- Improving referral and follow-up mechanisms, which involves applying engineering skills to develop and maintain reliable communication and transportation systems for healthcare coordination.
- The adaptation of simple technical options available to increase access to healthcare such as educative films and media outreach.

Overall, humanitarian engineering can help surgical oncology in Africa by addressing some of the barriers and challenges that limit the access and quality of cancer surgery in the region. Some of the ways that humanitarian engineering can help include the following:

- Developing and implementing low-cost and appropriate technologies and devices for cancer diagnosis, surgery, and postoperative care, such as portable ultrasound machines, surgical instruments, wound dressings, and pain management tools.
- Machine learning (artificial intelligence (AI)): this is the use of AI to analyze large amounts of data and extract patterns, insights, and predictions that can help in cancer diagnosis, prognosis, surgical oncology treatment planning, and monitoring. For instance, machine learning can be used to detect tumors from medical images, classify cancer types from molecular profiles, or optimize surgical oncology from patient data.
- Liquid biopsies and point-of-care diagnostics: these are methods that use blood or other bodily fluids to detect cancer biomarkers, such as circulating tumor cells, DNA, or proteins, without the need for invasive tissue biopsies. They can provide rapid, accurate, and low-cost diagnosis of cancer at the point of care, especially in remote or low-resource settings where access to conventional pathology services is limited.
- Low-cost surgical devices: these are devices that can perform surgical procedures for cancer treatment or palliation with minimal resources, such as electricity, water, or sterilization. For example, low-cost surgical devices can include battery-powered cautery tools, solar-powered autoclaves, or reusable surgical instruments.
- Improving and maintaining the infrastructure and equipment of health facilities, such as electricity, water, sanitation, sterilization, ventilation, and waste management systems.
- There is limited study about surgical oncologists in Africa and their perceptions of barriers to surgery. These data are important for stakeholders when designing and implementing surgical services in cancer management in LMICs like Africa.

- Training and supporting local health workers and surgeons in using and repairing the technologies and devices, as well as in improving their surgical skills and knowledge.
- Evaluating and monitoring the impact and sustainability of the humanitarian engineering interventions on the outcomes and quality of life of cancer patients
- Designing and implementing telemedicine and e-learning platforms to facilitate remote consultation, diagnosis, education, and training for cancer patients and health workers.
- Developing and testing innovative solutions to address the challenges of cancer surgery in low-resource settings, such as biodegradable implants, tissue engineering, and nanotechnology.
- Collaborating and partnering with local stakeholders, such as governments, NGOs, universities, hospitals, and communities, to ensure the relevance, feasibility, and sustainability of the humanitarian engineering interventions [5].

There have been some inspiring examples of how humanitarian engineering can contribute to humanitarian efforts on surgical oncology, providing innovative solutions to improve access, quality, and safety of surgical care in low-resource settings. Some examples are:

- Global Surgical Training Challenge, a project funded by the Bill & Melinda Gates Foundation and the Intuitive Foundation, that aims to create low-cost, open-source surgical training modules for LMICs.
- Surgical Innovation Fellowship, a program offered by the Massachusetts General Hospital and the Harvard Medical School, that trains surgeons to design and implement novel technologies and devices for surgical care in resource-limited settings.

Other examples that are benefiting oncology in low-resource settings include:

- Engineering World Health: this is an NGO that mobilizes biomedical engineering students and professionals to improve healthcare delivery in low-resource settings. The organization has supported the installation and maintenance of radiation therapy machines for cancer treatment in Rwanda, Tanzania, and Uganda. Similar efforts would be helpful in widening the reach of surgical oncology across the continent. Radiotherapy is in most cancer cases an important part of treatment with surgery.
- Engineers Without Borders USA: this is another nonprofit organization that empowers communities to meet their basic human needs through sustainable engineering projects. They have collaborated with the Rwanda Biomedical Center to design and construct a biosafety level 3 laboratory for cancer research and diagnosis in Kigali, Rwanda.

Some recommendations to advance humanitarian engineering in surgical oncology are:

- Anticipate and understand the local context and conduct a needs assessment before initiating any humanitarian engineering project or intervention, to ensure that the intervention is appropriate, feasible, and sustainable for the affected population.
- Incorporate best practices in the humanitarian delivery of surgery, such as following evidence-based guidelines, adhering to ethical principles, ensuring quality and safety standards, and monitoring and evaluating outcomes.
- Incorporate measures to promote local ownership, capacity building, and health system strengthening, such as engaging with local stakeholders, training local health workers, transferring skills and knowledge, and supporting infrastructure and equipment maintenance.
- Collaborate and coordinate with other humanitarian actors, such as the World Health Organization, the International Committee of the Red Cross, and other NGOs, to avoid duplication, share resources, and harmonize standards.
- Advocate for increased funding, research, and innovation in humanitarian engineering for surgical oncology, to address the unmet needs, gaps, and challenges in this field.

Humanitarian engineering in surgical oncology is definitely an area that could help overcome barriers in surgical oncology in LMICs and especially in Africa. Organizations like BCI and Peace and Love Hospitals are working to establish a surgical oncology center of excellence in Ghana that can serve the whole of West Africa and even serve as a surgical medical tourism hub for Africa. They are an example of organizations which are very open to collaborations including collaborative efforts in humanitarian engineering.

References

[1] Sullivan R, Alatise O I, Anderson B O, Audisio R, Autier P and Aggarwal A *et al* 2015 Global cancer surgery: delivering safe, affordable, and timely cancer surgery *Lancet Oncol.* **16** 1193–224

[2] Brand N R, Qu L G, Chao A and Ilbawi A M 2019 Delays and barriers to cancer care in low- and middle-income countries: a systematic review *Oncologist* **24** e1371–80

[3] McCutchan G, Weiss B, Quinn-Scoggins H, Dao A, Downs T and Deng Y *et al* 2021 Psychosocial influences on help-seeking behaviour for cancer in low-income and lower middle-income countries: a mixed-methods systematic review *BMJ Glob. Health* **6** e004213

[4] Ologunde R, Maruthappu M, Shanmugarajah K and Shalhoub J 2014 Surgical care in low and middle-income countries: burden and barriers *Int. J. Surg. Lond. Engl.* **12** 858–63

[5] Wiafe B A, Karekezi C, Jaklitsch M T and Audisio R A 2020 Perspectives on surgical oncology in Africa *Eur. J. Surg. Oncol.* **46** 3–5

IOP Publishing

Humanitarian Engineering for Global Oncology

Eric Ford

Chapter 7

Humanitarian engineering solutions: cervical cancer screening and treatment

Miriam Cremer, Rachel Masch, Montserrat Soler and Gabriel Conzuelo

7.1 Overview

Cervical cancer is a disease of inequality, as it remains a main cause of death in low- and middle-income countries (LMICs). This is a public health failure since primary (vaccination) and secondary (screening and treatment of precancer lesions) prevention methods exist. Tertiary protection is also essential, but surgical and radiation treatment, in addition to palliative care, are extremely limited in low-income settings. In recent years, there has been increasing momentum in developing innovations in secondary prevention strategies, as these are the most feasible and accessible in low-income settings. This chapter will focus on innovations and future directions in screening, early detection, and treatment of cervical precancerous lesions.

7.2 Cervical cancer worldwide

In high-resource settings, secondary prevention with cervical cytology (Pap smears) linked with timely treatment of precancer has led to sharp declines in cervical cancer mortality over the last century [1, 2]. However, the disease remains a main cause of death for women around the world, with nearly 90% of new cases diagnosed in LMICs [3]. The discovery that human papillomavirus (HPV) is the necessary cause of this disease has led to revolutionary prevention strategies such as the prophylactic HPV vaccination, which currently targets seven 'high-risk' HPV (hrHPV) types responsible for approximately 90% of cervical cancers worldwide [4–7]. There is promising evidence that a single dose of HPV vaccine yields protection against HPV 16 and HPV 18, the two types most frequently associated with development of invasive disease [8]. A single-dose vaccine would greatly reduce cost and facilitate uptake of vaccination programs. However, prophylactic HPV vaccines do not treat pre-existing HPV infections and related conditions [9, 10]. Thus, even if HPV vaccines were immediately disseminated worldwide, two to three generations of currently HPV-positive women would not benefit from them. Furthermore, the reality is that only a small percentage of girls and women are presently vaccinated worldwide, and access to vaccination in LMICs is extremely limited. Thus, cervical cancer mortality declines due to vaccination will not be seen for decades. To decrease the burden of disease, it remains necessary to increase access to screening and treatment of precancerous lesions in underserved populations.

7.3 Natural history of cervical cancer

Cervical cancer is the result of a long-term infection with one or more hrHPV subtypes. There are hundreds of HPV types, but only one is known to be an oncogenic or hrHPV types [11, 12]. Thus, HPV infection is a necessary condition for the development of cervical cancer [11, 12], although viral infection and behavioral determinants (e.g. smoking [13–15], sexual activity) are also risk factors [16]. Persistent hrHPV infection, on the other hand, can lead to premalignant cellular changes in the cervix known as cervical intraepithelial neoplasia (CIN). CIN is classified in grades 1–3. High-grade CIN (CIN2 or more severe, or CIN2+) is often used as an endpoint for clinical trials focused on screening and precancer treatment. While HPV infection and low-grade CIN are very common, especially in women under age 30, these generally clear spontaneously and do not lead to cervical cancer [17]. It is estimated that progression to malignancy occurs in about 1%, 5%, and 12% of CIN1, CIN2, and CIN3 cases, respectively [18–20]. Importantly, when it does progress, CIN takes between 10 and 15 years to develop into invasive cancer. Thus, there is a large window of opportunity for detecting and managing these lesions.

7.4 World Health Organization call to action

In 2018, the World Health Organization (WHO) called for action towards the global elimination of cervical cancer. A strategy for achieving this goal was ratified by member states in August 2020 and launched in November 2020 [21]. The WHO definition of elimination is a mortality rate of no more than 4 per 100 000 people. In 2020, the estimated cervical cancer mortality rate across all LMICs was 13.2 per 100 000 women (range 12.9–14.1/100 000) [22], but with some LMIC mortality estimates as high as 30 per 100 000 [23]. The targets of the WHO strategy are to vaccinate 90% of all girls between 9 and 15 years old, screen a minimum of 70% of women at age 35 and again at age 45, and provide treatment to 90% of women with underlying cervical disease. This is called the 90–70–90 strategy or 'triple-targets' [17, 22]. If vaccination and twice-in-a-lifetime cervical screening targets are met in all countries, 300 000 lives would be saved by 2030, 13.4 million lives by 2070, and 62 million lives by 2120 [17], a century from now. Low-cost, rapid, and effective innovations that increase access to screening and treatment of cervical precancer are necessary to speed up this timeline and avoid further unnecessary deaths.

7.5 Current cervical cancer screening strategies

7.5.1 Cytology (Pap smear)

The Pap smear has been the traditional method for cervical cancer screening since its invention in 1929 and is still widely used in many areas. It requires a multistep strategy of cytology (obtaining cells from the cervix), colposcopy (using magnification to identify abnormal areas on the cervix), laboratory analysis of samples, result delivery, and treatment of women diagnosed with cervical precancer. Despite large investments in cytology programs in low-resource settings, these have not resulted in a decrease in cervical cancer mortality [24]. Some of the barriers associated with cytology programs are lack of clinical and laboratory infrastructure [25], difficulty with patient follow-up [25–28], and the necessity of multiple screenings throughout a woman's lifetime.

7.5.2 Visual inspection with acetic acid

Visual inspection with acetic acid (VIA) uses 3%–5% diluted acetic acid applied to the cervix to turn cervical abnormalities white (acetowhite), which are then detectable by the naked eye. In some low-resource settings, VIA is an appropriate and effective approach to screen for cervical cancer [29, 30]. However, there is large variation in sensitivity for detection of CIN2+ [31, 32], and VIA programs have not made the impact that advocates had hoped for in reducing cervical disease burden. However, many LMICs continue to use VIA as a screening method because it yields immediate results, enables the provider to screen and treat precancerous lesions in the same visit, and does not require expensive equipment, specialized facilities, or highly trained personnel [29–36].

7.5.3 HPV diagnostic testing

Precancerous lesions caused by HPV infection take one to two decades to develop into invasive cancer [37]. Therefore, screening for HPV affords the opportunity to identify nearly all CIN2+ cases before progression [38]. Evidence has demonstrated that HPV testing has superior sensitivity to both cytology [39, 40] and VIA [40], although it has lower specificity [40]. Furthermore, there is strong evidence that HPV testing has an excellent negative predictive value, meaning that women who have a negative HPV test will be highly unlikely to have cervical cancer in the next 10 years [41–43]. An added advantage of many HPV tests is that unlike other screening methods that require a gynecological exam, women can easily self-collect the necessary sample at home or other nonclinical settings. Thus, there is wide consensus that in order ***to meet WHO elimination goals, it will be necessary for LMICs to move away from cytology and VIA towards HPV testing.*** This presents challenges but also opportunities for innovative biomedical engineering solutions that can increase the affordability and accessibility of HPV testing. Cervical cancer control strategies centered on HPV primary testing also raise important feasibility considerations that must be explored in tandem with triage and treatment alternatives that are necessary for effective prevention programs.

7.6 Challenges and innovations in cervical cancer screening

A successful cervical cancer prevention program must link widespread screening to safe and effective treatment. In addition, triage strategies that identify high-risk women must be put in place to minimize resource wastage and to avoid the undue burden of overtreatment on patients. In LMICs, implementation efforts are complicated by the need for affordability, the lack of highly trained personnel, and the accessibility barriers faced by many patients. Biomedical innovations in these areas must be developed with implementation strategies in mind in order to truly increase coverage in real-life settings. In the last decade, two such strategies have emerged:

- HPV self-sampling: while other screening strategies require a speculum exam by a qualified provider, HPV self-sampling bypasses this need and allows the woman to introduce a swab into her own vagina to collect cells for analysis. Although most of the data around self-collection come from high-income countries [44], self-sampling has been widely accepted among women in LMICs [45–47]. However, since collected cells in self-sampling come from the vagina rather than the cervix, the sensitivity of self-sampled HPV tests varies widely across manufacturers [48].
- Single-visit programs (e.g. 'screen-and-treat' or 'see-and-treat'): in settings where women face significant access barriers, screening with HPV testing or VIA followed by immediate treatment for screen-positive women increases coverage and reduces loss to follow-up [35, 49–51]. Importantly, self-collection can facilitate single-visit approaches. Each of these strategies requires accompanying technological innovations that make them feasible and effective in low-resource settings.

7.6.1 Innovations in HPV testing

HPV testing has emerged as a superior method to detect CIN2+ than both cytology and VIA, but challenges remain to facilitate widespread uptake in LMICs. A recent review revealed that there are over 250 HPV testing platforms currently on the market [52]. These HPV tests are often underregulated and operate on many different platforms [53], and most are targeted to high-income countries. The WHO provides regulatory prequalification for medicines and tests that meet unified standards of quality, safety, and efficacy. Since many LMICs rely on WHO prequalification to adopt clinical technologies, this is a crucial consideration in the development of HPV tests. As of September 2021, there are three HPV tests that have WHO prequalification: GeneXpert (Cepheid Inc, Sunnyvale, CA), RealTime (Abbott Laboratories, Chicago, IL), and careHPV (Qiagen, Gaithersburg, MD). All these tests have important limitations for use in LMICs (table 7.1). A promising new method is loop-mediated isothermal amplification (LAMP), a form of DNA amplification that does not require DNA extraction and is thus faster and lower-cost than other molecular

Table 7.1. WHO prequalified HPV tests currently on the market and the AmpFire test.

Platform	HPV detection/ self-sampling sensitivity	Requires batching	Process time	Cost per test under $10 USD	Other considerations for LMICs
Cepheid GeneXpert	Pooled result for any of 14 hrHPV/high	No	2 h	No	Large equipment, requires specialized laboratory, liquid buffer for transport
Abbott Real-Time PCR	HPV 16 and HPV 18 and pooled result for remaining 12 hrHPV/high	Yes	6 h	No	Requires specialized laboratory, liquid buffer for transport
Qiagen careHPV	Pooled result for any of 14 hrHPV/low	Yes	4–6 h	Yes	Requires electricity, hand pipetting, liquid buffer for transport
Modified AMPfire	By risk group: very high (16), high (18/45), medium (33/31/52/58/35), and low (39/51/59/56/68)/high (preliminary results)	No	1 h	Yes	Runs on any PCR machine, partly automated pipetting, dry samples for transport

testing methods. Atila BioSystems (Mountain View, CA) has developed AmpFire®, a test that utilizes LAMP to detect hrHPV types directly from clinical samples (that does not require liquid buffer required for collection, transport, and short-term storage) and can be run on any real-time polymerase chain reaction (PCR) machine. AmpFire obtained CE-mark (European regulatory approval) in 2017 but does not yet have WHO prequalification. A newly modified version reports HPV types by risk group classifications: 16, 18/45, 33/31/52/58/35, and 39/51/59/56/68 (see section XX) (table 7.1). The modified AmpFire® HPV screening test is cheaper, safer to transport, and easier to process than current prequalified tests; however large clinical trials are needed to verify effectiveness and regulatory approval needs to be obtained.

There is an urgent need for rapid, affordable, point-of-care HPV tests that can be deployed in low-income settings. The AmpFire® test has been designed to meet this need and may prove to be a turning point in cervical cancer prevention. Importantly, the equipment required to run the test is minimal, and the same platform can be used to process tests for other infectious agents, including SARS-COV-2.

7.7 Challenges and innovations in triage

While HPV positivity varies across populations, only between 1% and 2% of HPV-positive women are expected to have underlying CIN2+ [58, 59] (this percentage is significantly higher in populations with high rates of HIV infection [60–62], see section 7.8.4) or progress to invasive cancer. Thus, single-visit programs that treat all HPV-positive women significantly overtreat many women whose disease would spontaneously regress. In addition to few yet serious risks of treatment, this results in wastage of limited resources and places an unnecessary strain on women and the healthcare system. In high-resource settings, women with hrHPV undergo triage with colposcopy and biopsy to determine if high-grade precancer is present. In LMIC, triaging women with colposcopy is either not feasible or leads to a high loss to follow-up due to the requirement for multiple visits [26, 27]. Documented barriers in LMICs that limit adherence to programs with multiple visits include lack and cost of transportation, inability to leave work or find childcare, etc [63, 64]. A single-visit triage option is VIA, which consists of immediate visual evaluation after a positive HPV result. However, VIA has poor sensitivity [40] and it is highly subjective [65]. Alternative triage tests to cytology and VIA are needed to identify those at highest risk for invasive cancer immediately after a positive HPV result. Innovative alternatives include HPV genotyping, and imaging evaluation assisted by artificial intelligence.

7.7.1 Triage with HPV genotyping

The risk of developing cervical cancer depends on which HPV genotype is present. HPV genotyping tests can serve as both screening and triage tools that identify women at highest risk of developing invasive disease. There are 13 known types of HPV that are implicated in cervical cancer (other types have recently been discovered to be low risk) and these can be categorized into four ranked risk groups (table 7.2). Group 1 (HPV type 16) causes approximately 60% of all cervical cancer cases; Group 2 (HPV types 18 and 45) causes an additional 16.6%; Group 3 (HPV types 33, 31, 52, 58, and 35) each contribute between 2% and 4%; and Group 4 (HPV types 39, 51, 59, 56, and 68) have the lowest likelihood (individually from 0.6% to 1.6%) of being implicated in cervical cancer cases [12, 66]. Thus, knowing which underlying HPV type is present can be crucial for clinical management. In the United States, cervical cancer screening guidelines recently changed to include HPV genotyping as a management tool [67]. In LMICs however, this has not been feasible because of the high cost and complex infrastructure needed for genotyping HPV assays. The modified AmpFire® test is the only low-cost HPV genotyping test currently on the market, but large-scale trials are needed to determine its efficacy.

7.7.2 Triage with machine learning tools

The use of machine learning to identify high-grade disease using digital images is a promising avenue to incorporate triage strategies in screen-and-treat programs. These strategies involve taking a digital image of a cervix and running it through a

Table 7.2. Proposed 'extended' HPV genotyping for risk-based management.

HPV type	% of Cervical Cancers*	HPV types ranked by % of Cervical Cancers	9-year risk of progression to CIN3+ of incident HPV infection**	HPV types ranked by risk of CIN3+ of incident HPV infection	Proposed Risk Group
16	60.3	1	6.3	1	First
18	10.5	2	3.0	3	Second
45	6.1	3	2.2	5	Second
33	3.7	4	4.5	2	Third
31	3.6	5	2.2	5	Third
52	2.7	6	2.2	5	Third
58	2.2	7	1.9	6	Third
35	2.0	8	2.8	4	Third
39	1.6	9	1.1	7	Fourth
51	1.2	10	1.1	7	Fourth
59	1.1	11	0.9	9	Fourth
56	0.9	12	0.8	10	Fourth
68	0.6	13	1.0	8	Fourth

Demarco M. et al. EClinicalMed, 2020; De Sanjose et al. JNCI Can Spe, 2018

computer algorithm in order to determine the potential risk for cervical cancer according to its visual characteristics.

Currently, machine learning tools provide risk stratification based on colposcopic impression rather than colposcopic biopsy, the gold-standard diagnostic method. Colposcopic impression is known to be an insensitive measure of CIN2+, so it is unlikely that current algorithms have significant clinical utility. Stronger evidence comes from Automated Visual Evaluation (AVE), a prototype currently in development by a team of researchers at the National Cancer Institute. AVE is a machine learning image classifier that identifies cervical precancer from a digital colposcopic image captured with an ordinary smartphone. However, unlike other classifiers, AVE has been trained on biopsy-correlated images rather than colposcopic impression. AVE identifies patterns of high-grade precancer from cervical images and differentiates CIN2+ lesions from normal tissue and lower-grade dysplasia [68]. In clinical practice, AVE can be installed as a smartphone application and only requires the provider to be able to place a speculum and capture a clear image of the cervix. The magnified cervical image is then assessed and scored from 0.1 (lowest risk) to 1 (highest risk) of underlying CIN2. Retrospective preliminary data show that AVE has a higher detection rate of high-grade precancer than either cervicography (taking a picture of the cervix and sending it for expert review) or conventional cytology screening, and a similar performance to HPV testing and colposcopy for the detection of CIN2+.

7.8 Challenges and innovations in treatment

In some LMICs, up to 80% of women diagnosed with CIN2+ do not receive the recommended follow-up [26–28]. Thus, regardless of which screening strategy is used, effective cervical cancer prevention programs must link screening to adequate and affordable treatment [27]. Ideally, the range of therapies will include precancer treatment, treatment for patients with invasive disease, and palliative care at all stages of the disease. However, affordable or accessible treatment options for patients with invasive cervical cancer simply do not exist, and thus access to treatment for women with cervical cancer in LMICs is very limited. Most LMICs must rely on screening and precancer treatment as cervical cancer control methods. The WHO recommends two types of precancer treatment: excisional procedures that surgically remove the affected tissue and ablative treatments that utilize extreme temperatures to destroy precancerous cells.

Excisional procedures require expensive equipment, specialized training, and anesthesia. In contrast, ablative therapies are less expensive, do not require anesthesia [69], and can be administered by trained providers at all levels, including, physicians, nurses, and midwives [4, 6, 17–27]. For these reasons, cervical cancer programs in LMICs normally rely on ablation. The following is a summary of current cervical precancer treatment strategies, their challenges, and new solutions.

7.8.1 Excisional procedures (loop electrosurgical excision procedure and cold knife conization)

In high-income countries, loop electrosurgical excision procedure (LEEP) is the standard of care for treatment of cervical precancer. LEEP entails the use of a wire loop electrode to remove the entire cervical transformation zone. A Cochrane review reported cure rates of greater than 90% after LEEP. Some LMICs utilize cold knife conization (CKC), which utilizes a scalpel or laser knife to perform a similar procedure. In both cases, excised tissue is reviewed by a pathologist to ensure that the treatment is complete and that there is no invasive cancer. Despite high cure rates, there are several challenges to implementing LEEP or CKC in LMICs. These methods are more invasive than ablative methods and must be performed by highly trained medical personnel with technical expertise. In addition, LEEP and CKC require electricity, local anesthesia, pathology capacity, and resources for managing rare but serious adverse events. Wide use of LEEP is not realistic for low-resource settings. Yet, it is important that there is some LEEP capacity since ablation therapy is not appropriate for patients with large lesions, certain anatomic cervical variability, or suspicion of invasive disease [70, 71]. Such contraindications occur in about 10%–20% of HPV positive women. One manufacturer has developed a portable LEEP machine for low-resource settings (Liger Medical, Lehi, UT), but studies are needed to determine their efficacy and feasibility. Thus, there is still an unmet need for a safe excision technique that can be used in resource-poor and remote areas.

7.8.2 Cryotherapy

Cryotherapy is an ablative method that is the standard of care in many LMICs, and gas-based cryotherapy is by far the most common ablation technique used. A probe that has been cooled to very low temperatures with cryogenic gas is applied to cervical tissue, causing freezing and crystallization of the cervical intracellular water and producing necrosis (i.e. death) of the precancerous cells. Cryotherapy is a relatively inexpensive and simple procedure that can be performed by any level of healthcare worker with minimal training [72, 73]. It has also been found to be very safe, with low rates of immediate and long-term complications or long-term sequelae. The procedure is generally acceptable to women, and the most common side effects are mild cramping and watery discharge [72–74]. In general, ablative cryotherapy techniques have been shown to be 70%–90% effective in treating precancer [68–71, 79].

Limitations to gas-based cryotherapy: a cryotherapy unit cannot function without cryogenic gas (carbon dioxide or nitrous oxide), which needs to be purchased from a dispensary or a local bottling company. Gas tanks are not designed for portability, as their weight and size are substantial and cumbersome. The smallest gas tank weighs 18 pounds and can only be used to treat two patients. This severely limits large mobile health services that many LMICs rely on to treat women with limited access to health facilities. Procurement and cost are also barriers for gas-based cryotherapy. Nitrous oxide (N_2O) is considered the gold standard gas for the procedure, but it is not widely available in LMICs. When it can be readily sourced, it is significantly more expensive than carbon dioxide (CO_2). While CO_2 is generally more affordable and available than N_2O, its pricing, availability, and quality can be variable and unreliable.

Cryotherapy devices in development: the limitations posed by the need for gas have driven innovations in cryotherapy devices. The LMIC-adapted CryoPen® was specifically developed for use in low-resource settings. It uses a novel cryotherapy technology that does not require compressed gas and is powered by electricity or a car battery. The CryoPen reaches extremely cold temperatures with the use of a Stirling Cryocooler that chills the core to −105 °C. The CryoPen was originally designed to treat dermatologic conditions and was modified for gynecologic use in 2011. Although easier to transport than a cryogen gas tank, it is still cumbersome, weighing approximately 25 pounds. The device obtained FDA premarket approval, but the manufacturer pulled it from the market. Another device, the CryoPop, was developed by a research team from JHPIEGO and the Johns Hopkins Center for Bioengineering Innovation and Design. The CryoPop uses dry ice delivered in a simple, durable, and portable device that uses one-tenth of the CO_2 per procedure and costs half as much as existing devices. Necrosis has been found to be comparable to standard devices. CryoPop has CE mark. The major limitation to this device is that it still requires cryogenic gas. In recent years, cryotherapy solutions have been eclipsed by thermal ablation, as this method does not require gas and new devices have emphasized features that are important for use in LMICs.

7.8.3 Thermal ablation

Unlike cryotherapy, which uses cold temperatures to ablate tissue, thermal ablation (also known as cold coagulation, thermocoagulation) uses heat to destroy tissue. The superficial epithelium blisters off after treatment, and the underlying stroma and glandular crypts are destroyed by desiccation [76]. Although thermal ablation is not

Table 7.3. Thermal ablation devices available in the market.

Device	Power source	Cost	Treatments per charge of reusable battery	Other considerations for use in LMICs
WiSAP thermal ablator	Rechargeable battery or wall plug	2000 2 probes	120	Probe slider to avoid vaginal burns, high-level disinfectant
Liger Thermocoagulator	Rechargeable battery	1500 4 probes	30–60	Autoclave or high-level disinfection
Medgyn	Rechargeable battery	1550	100	Autoclave needed for sterilization

new, it was originally specifically developed for control of post-LEEP bleeding. Modifications of the existing desktop device (WiSAP Medical Technology GmbH, Brunnthal, Germany) and the creation of new portable, battery-operated devices (see table 7.3) have brought intense attention to this method as a game-changer in cervical cancer prevention. The WHO endorsed thermal ablation in 2020 as a treatment for CIN2+ [77]. This will increase its use in LMICs, but there are still important unanswered questions regarding variations in treatment protocol and long-term feasibility. A few studies have evaluated side effects and adverse events of thermal ablation [82–84]; Naud *et al* [83] using a sample of 52 women reported pain/cramps (79%) as the most common adverse event, followed by heat sensation in the vagina (25%), mild bleeding (2%), and vasovagal reaction (2%). Viviano *et al* [84] found that, while 95.5% of women ($n = 110$) experienced some degree of pain during therapy without anesthesia, the mean pain rating was mild or moderate. More data are needed on patient and provider acceptability.

Limitations of thermal ablation: clinicians in the United Kingdom have used thermal ablation for years to treat CIN2+ despite the paucity of rigorous clinical trial data on best practices. Retrospective evidence does exist in the literature, however, showing that thermal ablation has high efficacy and few side effects [78]. Much of the previous data, however, are based on the desktop device. Studies are currently ongoing to determine the optimal treatment protocol for portable devices in terms of temperature, duration, and the use of single or multiple probes.

7.8.4 Other treatment considerations

Obstetric outcomes of cervical precancer treatment: given the prevalence of ablative therapy in LMICs and the fact that many women treated for CIN2+ are of reproductive age, effects on obstetric outcomes are a significant concern. Overall, women with CIN are at greater risk for preterm birth and other perinatal morbidity compared to the general population [35]. While this risk tends to be higher for those who undergo excisional therapies, women treated with ablative methods are also susceptible to poor long-term obstetric outcomes [35]. Little research has been done on obstetric outcomes following ablative treatment, and most of these studies focus on any association with preterm birth [36–43]. Although there is some evidence that suggests fewer risks after ablative therapies, data on individual methods and comparisons among them are lacking.

HIV-infected women: several studies have reported that HIV-infected women are at greater risk of having multiple hrHPV types and persistent HPV infection associated with the development of CIN2+ compared to HIV-negative women [44–46]. There are no standard guidelines for cervical precancer treatment of HIV-infected women in LMICs, but some studies have examined the efficacy of ablative therapies in this population. The major methodological limitations concern case and outcome definition and assessment, which complicate the interpretation of cure rates. Although estimated cure rates are all above 75%, this proportion is lower than that estimated among HIV-uninfected women [9]. Follow-up is inconsistent across studies and attrition rates are high in many cases. Only one published study compared LEEP versus cryotherapy in a randomized controlled trial and found no difference in the 12-month cumulative incidence of CIN2+ [47]. There is a need for further research into ablative therapy for HIV-infected women. Studies in this area must clearly define cases and outcomes in accordance with international guidelines so they can the data can be easily reproduced. It is also important to consistently define follow-up time have at least 80% follow-up rate so that data will be more robust. It will also be necessary to define the best follow-up approach (e.g. return-visit versus video conference) and whether the existing screening protocols address the needs of this population. Comparisons between ablative therapies should also be conducted.

7.9 Future directions in cervical cancer prevention

7.9.1 p16/Ki-67

When HPV infections become cervical precancer the biomarkers p16 and Ki-67 are upregulated. Assays have been developed that can identify these biomarkers, and using these to triage HPV-positive women could help identify those at highest risk [36].

7.9.2 Host methylation

Another way to differentiate between an acute HPV infection and precancer/cancer is to evaluate the methylation of host genes. Increased methylation indicates precancer/cancer. Several genes have been studied as 'methylation markers'. Using these markers to triage people with positive HPV results has been shown to be similar to using cytology (Pap smears) [86, 87]. Advantages to using biomarkers for the triage of HPV-positive women include the objective evaluation of specimens of all kinds, including those that are self-collected, and the possibility of automating the process, allowing for a larger number of samples to be evaluated quickly. Novel automated processes for the analysis of methylated cells are being studied, including using magnets to move the methylated specimens through the different processing channels.

7.9.3 Viral methylation

HPV that is associated with precancer and cancer also has more viral methylation than the HPV in an acute infection across specific regions of the cell [88–90] and may also prove to be a good triage test for HPV-positive women.

7.9.4 E6/E7 tests

E6 and E7 are viral oncogenes that are expressed at a much higher rate in cervical precancers compared to transient HPV infections. The OncoE6 assay was developed to detect E6 in the highest HPV oncogenic subtypes (HPV 16/18/45), but although its specificity was good, its sensitivity to detect CIN3+ was lower than with the HPV-DNA tests. With further research and development, however, this could become a viable triage test, particularly in areas where histology is not available [91].

7.9.5 Urine HPV test

In recent years, urine detection of hrHPV has emerged as a potentially viable alternative for screening. There are two available versions of the test that evaluate specific areas in the viral genome: (1) Trovagene hrHPV test, targeting the E1 region and (2) OncoClarity BD urine HPV test, targeting the E6/E7 region. Both versions have achieved sensitivities of at least 80% in several studies [87–91]. However, specificity of the test is generally poor [87–90]. Presently, none of the urine-based tests have been approved by the FDA for clinical use. In the future, if these tests improve, they could increase access to cervical cancer screening, especially in people with limited access to healthcare.

7.10 Future directions in cervical cancer treatment

7.10.1 Therapeutic vaccines

As previously discussed, proteins E6 and E7 are significantly expressed in cervical precancers, but not elsewhere in the body. Furthermore, these proteins are necessary for cellular transformation and tumor progression. These two features make them ideal candidates for therapeutic vaccination, as they increase the host immune response against the tumor without damaging healthy tissue. In recent years, there have been numerous clinical trials aimed towards developing prophylactic vaccines for cervical cancer using different platforms. For instance, ongoing trials using peptide-, protein-, cell-, DNA-, RNA-, and vector-based vaccine candidates registered in the NIH ClinicalTrials.gov database. While some of these candidates have shown promising results and are currently in phases II and III, large heterogeneity across studies prevents comparisons between them [92]. Currently, none of these therapeutic vaccines have been approved by the FDA.

For the foreseeable future, secondary cervical cancer prevention in LMICs will primarily rely on a combination of screening and treatment of precancerous lesions. From a biomedical/engineering point of view, these procedures require the development of rapid, inexpensive, point-of-care devices that analyze samples with automated processes to minimize human error, and give results that are easy to interpret. The incorporation of triage tests in patients who are HPV positive will further ensure that the highest-risk women are receiving the treatment they need. Of course, HPV immunization also remains as vitally important as primary prevention, particularly the development of single-dose vaccination that limits loss to follow-up.

An often-quoted estimate of the average length of time between research and clinical practice is 17 years [93]. Although the reality is more complex, it is also the case that for LMICs, the lag may be even longer as most biomedical or engineering solutions are not developed with the limitations of low-resource settings in mind. To change this reality and close the massive gaps in global health disparities, it is essential to design products with 'implementation in mind', that is, emphasizing features that meet existing needs and that can rapidly and effectively move from the laboratory to the field. Engineering innovations that take into account the challenges of limited resources, poor infrastructure, access barriers for patients, and lack of trained personnel are much more likely to be adopted and sustained over time. The last decade has witnessed unprecedented interest in developing technologies to target cervical cancer morbidity and mortality. The momentum created by the WHO call to action can further sustain these efforts to achieving elimination of a cancer within our lifetime.

7.11 Conclusions

- Primary prevention for cervical can be achieved through HPV vaccines; however, vaccination alone is not sufficient to reduce mortality in the next century.
- Screening for HPV is the most sensitive way to detect precancer, but there is a great unmet need for rapid, affordable, point-of-care testing.
- Since not all HPV has oncogenic potential, genotyping can help differentiate the highest risk patients to save resources and avoid overtreatment.
- Screening must be linked to safe and effective precancer treatment and optimal implementation strategies such as self-sampling and single-visit approaches.
- The WHO has endorsed thermal ablation and this method is likely to be the future of precancer treatment, although an excision device that is suitable for low-resource settings is needed.
- Portable thermal ablation devices have been developed and there are active studies to determine best practices.
- Important unanswered questions regarding ablation treatment are long-term obstetric effects and efficacy in HIV-positive patients
- Future directions in cervical cancer prevention include point-of-care HPV testing, machine learning algorithms for testing and treatment, automated processing, effective triage stratigies and portable low-cost precancer treatment.
- Biomedical engineering innovations must take into account the implementation challenges of low-resource settings.

References

[1] Landis S H, Murray T, Bolden S and Wingo P A 1999 Cancer statistics, 1999 *CA: Cancer J. Clin.* **49** 8–31

[2] Wingo P A, Cardinez C J and Landis S H *et al* 2003 Long-term trends in cancer mortality in the United States, 1930–1998 *Cancer* **97** 3133–275

[3] Bray F, Ferlay J, Soerjomataram I, Siegel R L, Torre L A and Jemal A 2018 Global cancer statistics 2018: GLOBOCAN estimates of incidence and mortality worldwide for 36 cancers in 185 countries *CA: Cancer J. Clin.* **68** 394–424

[4] Blomberg M, Dehlendorff C, Sand C and Kjaer S K 2015 Dose-related differences in effectiveness of human papillomavirus vaccination against genital warts: a nationwide study of 550 000 young girls *Clin. Infect. Dis.* **61** 676–82

[5] Kang Y J, Lewis H and Smith M A *et al* 2015 Pre-vaccination type-specific HPV prevalence in confirmed cervical high grade lesions in the Māori and non-Māori populations in New Zealand *BMC Infect. Dis.* **15** 365

[6] Moore K A and Mehta V 2015 The growing epidemic of HPV-positive oropharyngeal carcinoma: a clinical review for primary care providers *J. Am. Board Fam. Med.* **28** 498–503

[7] Joura E A, Giuliano A R and Iversen O-E *et al* 2015 A 9-valent HPV vaccine against infection and intraepithelial neoplasia in women *New Engl. J. Med.* **372** 711–23

[8] Sankaranarayanan R, Joshi S and Muwonge R *et al* 2018 Can a single dose of human papillomavirus (HPV) vaccine prevent cervical cancer? Early findings from an Indian study *Vaccine* **36** 4783–91

[9] Hildesheim A, Herrero R and Wacholder S *et al* 2007 Effect of human papillomavirus 16/18 L1 viruslike particle vaccine among young women with preexisting infection: a randomized trial *JAMA* **298** 743–53

[10] 2007 Quadrivalent vaccine against human papillomavirus to prevent high-grade cervical lesions *New Engl. J. Med.* **356** 1915–27

[11] Okunade K S 2020 Human papillomavirus and cervical cancer *J. Obstetr. Gynaecol.* **40** 602–8

[12] Crosbie E J, Einstein M H, Franceschi S and Kitchener H C 2013 Human papillomavirus and cervical cancer *Lancet* **382** 889–99

[13] Plummer M, Herrero R and Franceschi S *et al* 2003 Smoking and cervical cancer: pooled analysis of the IARC multi-centric case-control study *Cancer Causes Control* **14** 805–14

[14] Kjellberg L, Hallmans G and Åhren A M *et al* 2000 Smoking, diet, pregnancy and oral contraceptive use as risk factors for cervical intra-epithelial neoplasia in relation to human papillomavirus infection *Br. J. Cancer* **82** 1332–8

[15] Roura E, Castellsagué X and Pawlita M *et al* 2014 Smoking as a major risk factor for cervical cancer and pre-cancer: results from the EPIC cohort *Int. J. Cancer* **135** 453–66

[16] Doorbar J, Quint W and Banks L *et al* 2012 The biology and life-cycle of human papillomaviruses *Vaccine* **30** F55–70

[17] Brisson M, Kim J J and Canfell K *et al* 2020 Impact of HPV vaccination and cervical screening on cervical cancer elimination: a comparative modelling analysis in 78 low-income and lower-middle-income countries *Lancet* **395** 575–90

[18] ÖStör A G 1993 Natural history of cervical intraepithelial neoplasia: a critical review *Int. J. Gynecol. Pathol.* **12** 182–92

[19] McCredie M R, Sharples K J and Paul C *et al* 2008 Natural history of cervical neoplasia and risk of invasive cancer in women with cervical intraepithelial neoplasia 3: a retrospective cohort study *Lancet Oncol.* **9** 425–34

[20] Massad L S, Evans C T, Minkoff H and Passaro D J 2004 Natural history of grade 1 cervical intraepithelial neoplasia in women with human immunodeficiency virus *Obstetr. Gynecol.* **104** 1077–85

[21] World Health Organization 2018 *WHO Director-General Calls for All Countries to Take Action to Help End the Suffering Caused by Cervical Cancer* (https://who.int/reproductive-health/call-to-action-elimination-cervical-cancer/en/) (accessed 29 August 2021)

[22] Canfell K, Kim J J and Brisson M *et al* 2020 Mortality impact of achieving WHO cervical cancer elimination targets: a comparative modelling analysis in 78 low-income and lower-middle-income countries *Lancet* **395** 591–603

[23] LaVigne A W, Triedman S A, Randall T C, Trimble E L and Viswanathan A N 2017 Cervical cancer in low and middle income countries: addressing barriers to radiotherapy delivery *Gynecol. Oncol. Rep.* **22** 16–20

[24] Catarino R, Petignat P, Dongui G and Vassilakos P 2015 Cervical cancer screening in developing countries at a crossroad: emerging technologies and policy choices *World J. Clin. Oncol.* **6** 281–90

[25] Agurto I, Sandoval J, de La Rosa M and Guardado M E 2006 Improving cervical cancer prevention in a developing country *Int. J. Qual. Health Care* **18** 81–6

[26] Maza M, Matesanz S and Alfaro K *et al* 2016 Adherence to recommended follow-up care after high-grade cytology in El Salvador *Int. J. Healthcare* **2** 31–6

[27] Gage J C, Ferreccio C, Gonzales M, Arroyo R, Huivín M and Robles S C 2003 Follow-up care of women with an abnormal cytology in a low-resource setting *Cancer Detect. Prev.* **27** 466–71

[28] Jeong S J, Saroha E, Knight J, Roofe M and Jolly P E 2011 Determinants of adequate follow-up of an abnormal Papanicolaou result among Jamaican women in Portland, Jamaica *Cancer Epidemiol.* **35** 211–6

[29] Sauvaget C, Fayette J-M, Muwonge R, Wesley R and Sankaranarayanan R 2011 Accuracy of visual inspection with acetic acid for cervical cancer screening *Int. J. Gynecol. Obstetr.* **113** 14–24

[30] Sankaranarayanan R, Esmy P O and Rajkumar R *et al* 2007 Effect of visual screening on cervical cancer incidence and mortality in Tamil Nadu, India: a cluster-randomised trial *Lancet* **370** 398–406

[31] Basu P, Mittal S and Banerjee D *et al* 2015 Diagnostic accuracy of VIA and HPV detection as primary and sequential screening tests in a cervical cancer screening demonstration project in India *Int. J. Cancer* **137** 859–67

[32] Cuzick J, Arbyn M and Sankaranarayanan R *et al* 2008 Overview of human papillomavirus-based and other novel options for cervical cancer screening in developed and developing countries *Vaccine* **26** K29–41

[33] Blumenthal P D, Gaffikin L, Chirenje Z M, McGrath J, Womack S and Shah K 2001 Adjunctive testing for cervical cancer in low resource settings with visual inspection, HPV, and the Pap smear *Int. J. Gynecol. Obstetr.* **72** 47–53

[34] Gaffikin L, Blumenthal P D and Emerso M *et al* 2003 Safety, acceptability, and feasibility of a single-visit approach to cervical-cancer prevention in rural Thailand: a demonstration project *Lancet* **361** 814–20

[35] Sankaranarayanan R, Rajkumar R and Esmy P O *et al* 2007 Effectiveness, safety and acceptability of 'see and treat' with cryotherapy by nurses in a cervical screening study in India *Br. J. Cancer* **96** 738–43

[36] Wentzensen N, Schiffman M, Palmer T and Arbyn M 2016 Triage of HPV positive women in cervical cancer screening *J. Clin. Virol.* **76** S49–55

[37] Tota J E, Chevarie-Davis M, Richardson L A, deVries M and Franco E L 2011 Epidemiology and burden of HPV infection and related diseases: implications for prevention strategies *Prevent. Med.* **53** S12–21

[38] Gage J C, Schiffman M and Katki H A *et al* 2014 Reassurance against future risk of precancer and cancer conferred by a negative human papillomavirus test *JNCI J. Natl. Cancer Inst.* **106** dju153

[39] Kulasingam S L, Hughes J P and Kiviat N B *et al* 2002 Evaluation of human papillomavirus testing in primary screening for cervical abnormalities *JAMA* **288** 1749–57

[40] Longatto-Filho A, Naud P and Derchain S F M *et al* 2012 Performance characteristics of Pap test, VIA, VILI, HR-HPV testing, cervicography, and colposcopy in diagnosis of significant cervical pathology *Virchows Arch.* **460** 577–85

[41] Kjaer S, Høgdall E and Frederiksen K *et al* 2006 The absolute risk of cervical abnormalities in high-risk human papillomavirus–positive, cytologically normal women over a 10-year period *Cancer Res.* **66** 10630–6

[42] Schiffman M, Glass A G and Wentzensen N *et al* 2011 A long-term prospective study of type-specific human papillomavirus infection and risk of cervical neoplasia among 20,000 women in the Portland Kaiser cohort study *Cancer Epidemiol. Biomarkers Prev.* **20** 1398–409

[43] Elfström K M, Smelov V and Johansson A L V *et al* 2014 Long term duration of protective effect for HPV negative women: follow-up of primary HPV screening randomised controlled trial *BMJ (Online)* **348** g130

[44] Yeh P T, Kennedy C E, de Vuyst H and Narasimhan M 2019 Self-sampling for human papillomavirus (HPV) testing: a systematic review and meta-analysis *BMJ Global Health* **4** e001351

[45] Lack N, West B and Jeffries D *et al* 2005 Comparison of non-invasive sampling methods for detection of HPV in rural African women *Sex. Transm. Infect.* **81** 239–41

[46] Modibbo F, Iregbu K C and Okuma J *et al* 2017 Randomized trial evaluating self-sampling for HPV DNA based tests for cervical cancer screening in Nigeria *Infect. Agents Cancer* **12** 11

[47] Viviano M, Tran P L and Kenfack B *et al* 2018 Self-versus physician-collected samples for the follow-up of human papillomavirus-positive women in sub-Saharan Africa *Int. J. Women's Health* **10** 187–94

[48] Hawkes D, Keung M H T and Huang Y *et al* 2020 Self-collection for cervical screening programs: from research to reality *Cancers* **12** 1053

[49] Kunckler M, Schumacher F and Kenfack B *et al* 2017 Cervical cancer screening in a low-resource setting: a pilot study on an HPV-based screen-and-treat approach *Cancer Med.* **6** 1752–61

[50] Alfaro K, Maza M and Felix J C *et al* 2020 Outcomes for step-wise implementation of a human papillomavirus testing-based cervical screen-and-treat program in El Salvador *JCO Global Oncol.* **6** 1519–30

[51] Martin C E, Tergas A I, Wysong M, Reinsel M, Estep D and Varallo J 2014 Evaluation of a single-visit approach to cervical cancer screening and treatment in Guyana: feasibility, effectiveness and lessons learned *J. Obstetr. Gynaecol. Res.* **40** 1707–16

[52] Poljak M, Kocjan B J, Oštrbenk A and Seme K 2016 Commercially available molecular tests for human papillomaviruses (HPV): 2015 update *J. Clin. Virol.* **76** S3–13

[53] Cubie H A and Campbell C 2020 Cervical cancer screening—the challenges of complete pathways of care in low-income countries: focus on Malawi *Women's Health* **16** 1745506520914804
[54] Huang S, Tang N and Mak W B *et al* 2009 Principles and analytical performance of Abbott RealTime High Risk HPV test *J. Clin. Virol.* **45** S13–7
[55] Trope L A, Chumworathayi B and Blumenthal P D 2013 Feasibility of community-based careHPV for cervical cancer prevention in rural Thailand *J. Low. Genit. Tract Dis.* **17** 315–9
[56] Cholli P, Bradford L and Manga S *et al* 2018 Screening for cervical cancer among HIV-positive and HIV-negative women in Cameroon using simultaneous co-testing with careHPV DNA testing and visual inspection enhanced by digital cervicography: findings of initial screening and one-year follow-up *Gynecol. Oncol.* **148** 118–25
[57] Goldstein A, Goldstein L S and Lipson R *et al* 2020 Assessing the feasibility of a rapid, high-volume cervical cancer screening programme using HPV self-sampling and digital colposcopy in rural regions of Yunnan, China *BMJ Open* **10** e035153
[58] Koliopoulos G, Nyaga V N and Santesso N *et al* 2017 Cytology versus HPV testing for cervical cancer screening in the general population *Cochrane Database Syst. Rev.* **2017** CD008587
[59] Qiao Y L, Zhao F H and Lewkowitz A K *et al* 2012 Prevalence of human papillomavirus and cervical intraepithelial neoplasia in China: a pooled analysis of 17 population-based studies *Int. J. Cancer* **131** 2929–38
[60] Zhang H Y, Tiggelaar S M and Sahasrabuddhe V v *et al* 2012 HPV prevalence and cervical intraepithelial neoplasia among HIV-infected women in Yunnan Province, China: a pilot study *Asian Pac. J. Cancer Prev.* **13** 91–6
[61] de Vuyst H, Mugo N R and Chung M H *et al* 2012 Prevalence and determinants of human papillomavirus infection and cervical lesions in HIV-positive women in Kenya *Br. J. Cancer* **107** 1624–30
[62] Stewart Massad L, Xie X and D'Souza G *et al* 2015 Incidence of cervical precancers among HIV-seropositive women *Am. J. Obstetr. Gynecol.* **212** 606.e1–606.e8
[63] Paz-Soldán V A, Bayer A M, Nussbaum L and Cabrera L 2012 Structural barriers to screening for and treatment of cervical cancer in Peru *Reprod. Health Matters* **20** 49–58
[64] Austad K, Chary A and Xocop S M *et al* 2018 Barriers to cervical cancer screening and the cervical cancer care continuum in rural Guatemala: a mixed-method analysis *J. Global Oncol.* **2018** 1–10
[65] Arab M, Moridi A and Fazli G *et al* 2021 Is visual inspection with acetic acid (VIA) a useful method of finding pre-invasive cervical cancer? *Clin. Exp. Obstetr. Gynecol.* **48** 128–31
[66] Burd E M 2003 Human papillomavirus and cervical cancer *Clin. Microbiol. Rev.* **16** 1–17
[67] Cheung L C, Egemen D and Chen X *et al* 2020 2019 ASCCP risk-based management consensus guidelines: methods for risk estimation, recommended management, and validation *J. Low. Genit. Tract Dis.* **24** 90–101
[68] Pal A, Xue Z and Befano B *et al* 2021 Deep metric learning for cervical image classification *IEEE Access* **9** 53266–75
[69] Duncan I D, McKinley C A, Pinion S B and Wilson S M 2005 A double-blind, randomized, placebo-controlled trial of prilocaine and felypressin (Citanest and Octapressin) for the relief of pain associated with cervical biopsy and treatment with the Semm coagulator *J. Low. Genit. Tract Dis.* **9** 171–5

[70] Guidelines Review Committee 2019 *WHO Guidelines for the Use of Thermal Ablation for Cervical Pre-Cancer Lesions Who.int* (https://who.int/publications/i/item/9789241550598) (Accessed 29 August 2021)
[71] World Health Organization 2013 *WHO Guidelines for Screening and Treatment of Precancerous Lesions for Cervical Cancer Prevention* (https://apps.who.int/iris/bitstream/handle/10665/94830/9789241548694_eng.pdf?sequence=1) (Accessed 29 August 2021)
[72] Adefuye P O, Dada O A, Adefuye B O, Shorunmu T O, Akinyemi B O and Idowu-Ajiboye B O 2015 Feasibility, acceptability, and effectiveness of visual inspection of the cervix with acetic acid and cryotherapy for dysplasia in Nigeria *Int. J. Gynecol. Obstetr.* **129** 62–6
[73] Phongsavan K, Phengsavanh A, Wahlström R and Marions L 2011 Safety, feasibility, and acceptability of visual inspection with acetic acid and immediate treatment with cryotherapy in rural Laos *Int. J. Gynecol. Obstetr.* **114** 268–72
[74] Fallala M S and Mash R 2015 Cervical cancer screening: safety, acceptability, and feasibility of a single-visit approach in Bulawayo, Zimbabwe *Afr. J. Prim. Health Care Fam. Med.* **7** 742
[75] Santesso N, Schünemann H and Blumenthal P *et al* 2012 World Health Organization Guidelines: use of cryotherapy for cervical intraepithelial neoplasia *Int. J. Gynecol. Obstetr.* **118** 97–102
[76] Martin-Hirsch P P L, Paraskevaidis E, Bryant A and Dickinson H O 2013 Surgery for cervical intraepithelial neoplasia *Cochrane Database Syst. Rev.* **2013** CD001318
[77] Mcclung E C and Blumenthal P D 2012 Efficacy, safety, acceptability and affordability of cryotherapy: a review of current literature *Minerva Ginecol.* **64** 149–71
[78] Sauvaget C, Muwonge R and Sankaranarayanan R 2013 Meta-analysis of the effectiveness of cryotherapy in the treatment of cervical intraepithelial neoplasia *Int. J. Gynecol. Obstetr.* **120** 218–23
[79] D'Alessandro P, Arduino B and Borgo M *et al* 2018 Loop electrosurgical excision procedure versus cryotherapy in the treatment of cervical intraepithelialneoplasia: a systematic review and meta-analysis of randomized controlled trials *Gynecol. Minim. Invasive Ther.* **7** 145–51
[80] Semm K 1966 New apparatus for the 'cold-coagulation' of benign cervical lesions *Am. J. Obstetr. Gynecol.* **95** 963–6
[81] World Health Organization (WHO) 2021 *WHO Guideline for Screening and Treatment of Cervical Pre-Cancer Lesions for Cervical Cancer Prevention* 2nd edn (Geneva: World Health Organization)
[82] Campbell C, Kafwafwa S and Brown H *et al* 2016 Use of thermo-coagulation as an alternative treatment modality in a 'screen-and-treat' programme of cervical screening in rural Malawi *Int. J. Cancer* **139** 908–15
[83] Naud P S V, Muwonge R, Passos E P, Magno V, Matos J and Sankaranarayanan R 2016 Efficacy, safety, and acceptability of thermocoagulation for treatment of cervical intra-epithelial neoplasia in a hospital setting in Brazil *Int. J. Gynecol. Obstetr.* **133** 351–4
[84] Viviano M, Kenfack B and Catarino R *et al* 2017 Feasibility of thermocoagulation in a screen-and-treat approach for the treatment of cervical precancerous lesions in sub-Saharan Africa *BMC Women's Health* **17** 2
[85] Dolman L, Sauvaget C, Muwonge R and Sankaranarayanan R 2014 Meta-analysis of the efficacy of cold coagulation as a treatment method for cervical intraepithelial neoplasia: a systematic review *BJOG: Int. J. Obstetr. Gynaecol.* **121** 929–42

[86] Derbie A, Mekonnen D, Woldeamanuel Y, van Ostade X and Abebe T 2020 HPV E6/E7 mRNA test for the detection of high grade cervical intraepithelial neoplasia (CIN2+): a systematic review *Infect. Agents Cancer* **15** 9

[87] Lorenzi A T, Fregnani J H T and Dockter J *et al* 2018 High-risk human papillomavirus detection in urine samples from a referral population with cervical biopsy-proven high-grade lesions *J. Low. Genit. Tract Dis.* **22** 17–20

[88] Cuzick J, Cadman L and Ahmad A S *et al* 2017 Performance and diagnostic accuracy of a urine-based human papillomavirus assay in a referral population *Cancer Epidemiol. Biomarkers Prev.* **26** 1053–9

[89] Sahasrabuddhe V v, Gravitt P E and Dunn S T *et al* 2014 Evaluation of clinical performance of a novel urine-based HPV detection assay among women attending a colposcopy clinic *J. Clin. Virol.* **60** 414–7

[90] Rohner E, Rahangdale L and Sanusi B *et al* 2020 Test accuracy of human papillomavirus in urine for detection of cervical intraepithelial neoplasia *J. Clin. Microbiol.* **58** e01443–e19

[91] Leeman A, del Pino M and Molijn A *et al* 2017 HPV testing in first-void urine provides sensitivity for CIN2+ detection comparable with a smear taken by a clinician or a brush-based self-sample: cross-sectional data from a triage population *BJOG: Int. J. Obstetr. Gynaecol.* **124** 1356–63

[92] Smalley Rumfield C, Roller N, Pellom S T, Schlom J and Jochems C 2020 Therapeutic vaccines for HPV-associated malignancies *ImmunoTargets Ther.* **9** 167–200

[93] Morris Z S, Wooding S and Grant J 2011 The answer is 17 years, what is the question: understanding time lags in translational research *J. R. Soc. Med.* **104** 510–20

IOP Publishing

Humanitarian Engineering for Global Oncology

Eric Ford

Chapter 8

Humanitarian engineering solutions: medical oncology

Jicheng Zhang, Jiangbo Yu, Jerald P Radich and Daniel T Chiu

8.1 Introduction: reducing the cost and complexity of diagnostic assays by using polymer dots

Fluorescence assays are ubiquitous in diagnostics in oncology, immunology, and other disease-related fields, and include polymerase chain reaction and other nucleic acid assays, Western blot and other protein assays, and flow cytometry and microfluidics-based cell assays. However, a large fraction of current fluorescence assays are poorly suited for use in low- and middle-income countries (LMICs) because they require expensive and complex instrumentation and extensive training. One strategy to make fluorescence assays more available for use in LMICs is to redesign the instrumentation to simplify their use and reduce their cost. An alternate strategy is to make the fluorescent probes used in the assays significantly brighter so that complex, expensive systems with laser excitation, expensive optics, ultra-sensitive cameras, desktop computers, and large footprints can be replaced with small, portable, inexpensive LEDs or ultraviolet (UV) lamps and smartphones. Semiconducting polymer dots (Pdots) enable the latter strategy. Pdots are fluorescent nanoparticles (5–30 nm diameter) composed of well-packed π-conjugated fluorescent semiconducting polymers, and possess much greater single-particle brightness than quantum dots (Qdots) or small-molecule fluorophores. In addition to being extremely bright, Pdots are photostable and chemically stable [1], have a low cytotoxicity suitable for *in vivo* assays [2], can be manufactured cheaply and reproducibly [3], and can be customized for a wide range of diagnostic and theranostic applications by varying their optical properties via polymer composition and their bioreactivity via bioconjugation [4].

In this chapter we define Pdots and describe their physicochemical and optical properties, including their high brightness and how their optical properties can be tuned via polymer composition to further increase assay signal-to-noise ratio

doi:10.1088/978-0-7503-3751-9ch8

(e.g. by using near-infrared (NIR)-emitting or long-timescale-emitting Pdots for avoidance of tissue autofluorescence *in vivo*, and by using dual-wavelength-emissive Pdots that allow ratiometric measurements and internally-calibrated assays) (section 8.2); we describe how Pdots are prepared and can be modified for different applications through bioconjugation (section 8.3), and we provide examples of applications of Pdots to oncology including diagnostic assays to detect cancer biomarkers, tumor imaging, and theranostics such as photodynamic therapy (section 8.4).

8.2 What are Pdots? Physicochemical and optical properties

Pdots are semiconducting polymer nanoparticles that are composed primarily of π-conjugated polymers—polymers with a backbone of alternating double- and single-bonds. Pdots are distinguished from other conjugated polymer nanoparticles by their small size (5–30 nm) (hence the 'dots' in their name), their high brightness, and their composition which by definition must include more than 50% semiconducting polymer, and must contain a hydrophobic polymer interior to allow the high semiconducting polymer packing density required to achieve their small size and high brightness [5].

The fluorescent semiconducting polymers used to create Pdots include polyfluorenes, poly(phenylene ethynylene), poly(phenylene vinylene) copolymers, and polyfluorene copolymers (figure 8.1(A)). These semiconducting polymers have been designed to exhibit an array of emission colors across the visible spectrum (figures 8.1(B),(D)) [6]. The absorption spectra of Pdots formed using these polymers range from 350–600 nm (figure 8.1(C)), and are usually broader and blue-shifted relative to those of their free polymers in tetrahydrofuran (THF) solution [7]. This behavior should be caused by a reduction in conjugation length, resulting from the bending, torsion, and kinking of the polymer backbone.

Pdots exhibit advantages over small-molecule fluorescent probes and other fluorescent nanoparticles like Qdots, with their brightness exceeding that of these counterparts by an order of magnitude or more. This enhancement can be attributed primarily to their significantly larger absorption cross section. Fluorescence brightness can be accurately estimated by the product of two key factors: the peak absorption cross section and the fluorescence quantum yield. The absorption cross section reflects the light-harvesting capability of a fluorescent particle at a specific wavelength, and this parameter can be obtained from the particle's absorption spectra. The peak absorption cross section of ~15 nm diameter Pdots is $\sim 10^{-13}$ cm^2, which is 10–100 times as large as that of cadmium selenide Qdots in the visible and near-UV range. Fluorescence quantum yield serves as a measure of the efficiency of a fluorescent probe to emit light, representing the ratio of emitted photons to absorbed photons. Pdots with high fluorescence quantum yields ranging from 50%–60% have been reported [9, 10]. These values are comparable to the quantum yields of inorganic Qdots. In order to evaluate the single-particle brightness of Pdots in relation to other fluorophores, we conducted a comparative study using the well-known poly[(9,9-dioctylfluorenyl-2,7-diyl)-co-(1,4-benzo-{2,1,3}-thiadiazole) (PFBT) Pdots and two commonly utilized fluorescent probes, namely Qdot 565 and IgG Alexa

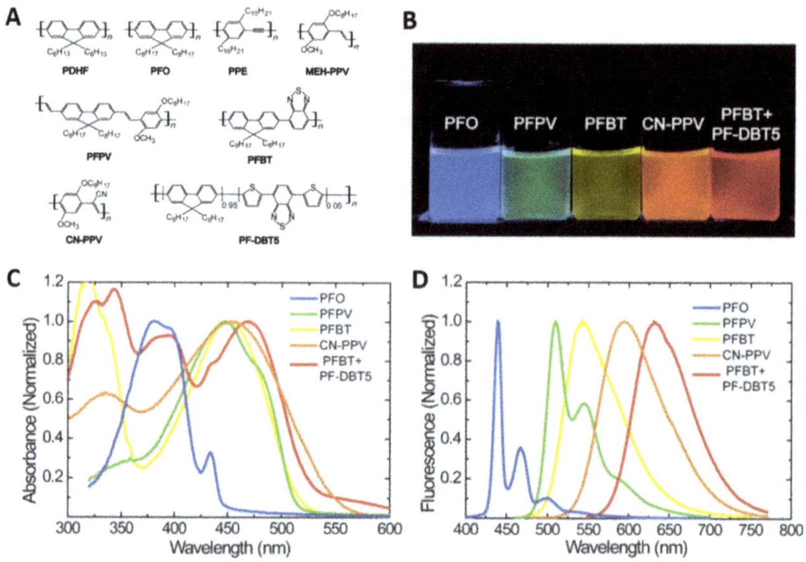

Figure 8.1. Fluorescent semiconducting polymers and corresponding Pdot absorption and emission properties. (A) Chemical structures of fluorescent semiconducting polymers used to make Pdots. (B) Pdots formed using these polymers have emission colors that span the visible spectrum. (C) Pdot absorption spectra. (D) Pdot emission spectra. Abbreviations: PDHF, poly(9,9-dihexylfluorenyl-2,7-diyl); PFO, poly(9,9-dioctylfluorenyl-2,7-diyl); PPE, poly(phenylene ethynylene); MEH-PPV, poly[2-methoxy-5-(2-ethylhexyloxy)-1,4-phenylenevinylene]; CN-PPV, poly[2-methoxy-5-(2-ethylhexyloxy)-1,4-(1-cyanovinylene-1,4-phenylene)]; PFPV, poly [{9,9-dioctyl-2,7-divinylene-fluorenylene}-alt-co-{2-methoxy-5-(2-ethylhexyloxy)-1,4-phenylene}]; PFBT, poly [(9,9-dioctylfluorenyl-2,7-diyl)-co-(1,4-benzo-{2,1,3}-thiadiazole); PF-DBT5, poly(9,9-dioctylfluorene)-co-(4,7-di-2-thienyl-2,1,3-benzothiadiazole). Pdot spectra adapted with permission from [8]. Copyright (2008) American Chemical Society. Also adapted from [9] John Wiley & Sons, Copyright (2011).

Fluor 488 (approximately six dye molecules per IgG) [11]. Under laser excitation at 488 nm, The brightness of ~10–15 nm PFBT Pdots surpassed that of IgG Alexa Fluor 488 and Qdot 565 by approximately 30-fold, as evidenced by the fluorescence intensity distribution of several thousands of particles (figure 8.2(A)). This enhancement agrees with the calculations based on photophysical parameters. In a similar comparison of single-particle brightness [9], polymer-blend Pdots comprising PFBT and the red-emissive polymer poly(9,9-dioctylfluorene)-co-(4,7-di-2-thienyl-2,1,3-benzothiadiazole) (PF-DBT5) exhibited brightness approximately 15 times greater than a Qdot emitting at 655 nm (figure 8.2(B)).

In addition to being brighter, Pdots can be made more photostable than other fluorescent nanoparticles such as Qdots, with over 10-fold slower single-particle photobleaching. Fluorescent molecules or nanoparticles' photostability is characterized by the photobleaching quantum yield (ϕ_B). This metric quantifies the proportion of photobleached molecules relative to the total number of absorbed photons over a defined time interval. (A smaller ϕ_B corresponds to greater photostability.) Typical single-molecule fluorescent dyes exhibit ϕ_B in the range of 10^{-4} to 10^{-6} [12]. Photobleaching kinetics of aqueous Pdot suspensions vary substantially

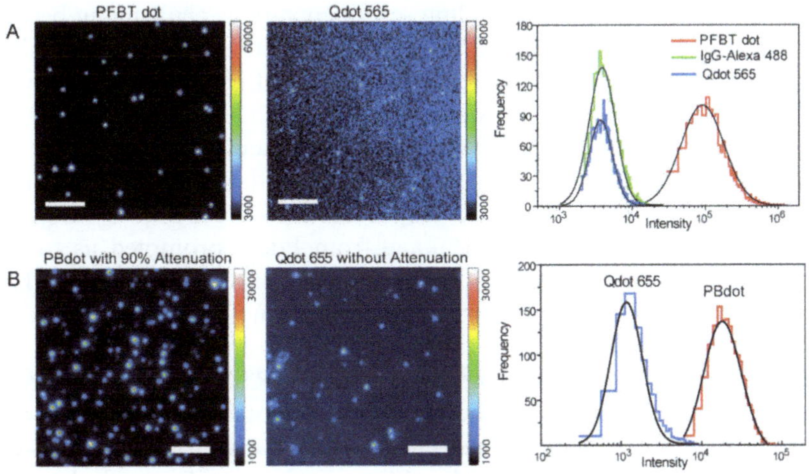

Figure 8.2. Pdots are more than an order of magnitude brighter than Qdots and small-molecule fluorophores. (A) Single-particle fluorescence images of PFBT Pdots and Qdot 565 acquired using identical 488-nm excitation and detection settings. Scale bar, 5 mm. The right panel illustrates intensity distributions derived from single-particle fluorescence data, and includes IgG-Alexa Fluor 488 conjugates containing about six dyes per conjugate. Reproduced from [11] with permission. (B) Single-particle fluorescence images of a polymer-blend Pdot ('PBdot') consisting of PFBT and PF-DBT5 and Qdot 655. Excitation was identical (488 nm); a neutral density filter with an optical density of 1 was used to collect fluorescence images for PBdots but not for Qdot 655. Scale bar, 4 mm. The right panel displays intensity distributions obtained from single-particle fluorescence data. Adapted from [9] John Wiley & Sons, Copyright (2011).

depending on the polymer, with ϕ_B ranging from 10^{-7} to 10^{-10} based on rate constants from photobleaching kinetics measurements [8, 11]. Through the division of the fluorescence quantum yield by the photobleaching quantum yield, we can derive the 'photon number', representing the total emitted photons before photobleaching. For instance, the photon number for 15 nm PFBT dots was estimated to be 10^9, three to four orders of magnitude higher than those typically observed for single-molecule fluorophores [8, 11]. Additional information about photostability is derived from single-particle photobleaching studies. Statistical analysis of photobleaching trajectories indicated $\sim 6 \times 10^8$ photons emitted per ~10 nm Pdot particle with some particles emitting over 10^9 photons, whereas Qdots emitted $\sim 10^7$ photons per particle [13].

The optical properties of Pdots can be finely tuned by adjusting their polymer composition to further improve assay signal-to-noise ratio and reduce instrumentation sensitivity requirements. Three examples, described below, are (1) NIR-emissive Pdots that increase signal-to-noise ratio by taking advantage of the greater penetration of NIR light through biological tissues and the greater spectral separation of NIR light from autofluorescence than for other emission wavelengths; (2) Pdots with time-delayed emission to minimize background by time-gating out autofluorescence, which has a short emission lifetime; and (3) Pdots with stimuli-responsive emissions at two different wavelengths that allow ratiometric measurements for robust, internally-calibrated assays.

In developing NIR emitters, avoiding self-quenching is challenging because NIR emitters possess extended π-conjugations which results in decreased solubility and increased aggregation in aqueous solutions. Pdots offer a partial solution to this issue, thanks to their hydrophobic interior, which provides an excellent environment for hosting NIR chromophores. This characteristic helps to mitigate the solubility and aggregation challenges typically associated with NIR emitters in aqueous solutions. We first attempted to develop NIR-emissive Pdots by doping Pdots with NIR dyes [14]; however, dye leakage from Pdots prompted us to employ a different strategy. We next designed Pdots which incorporated BODIPY-based and squaraine-based NIR emitters as part of the semiconducting polymer backbone [15]. We created two different highly luminescent ~15 nm NIR-emissive Pdots by polymerizing Förster resonance energy transfer (FRET) donors, acceptors, and energy transfer units into the backbone of the semiconducting polymer, and using the resulting polymers to form Pdots (figures 8.3(A) and (B)). The Pdots exhibited narrow fluorescence bands (FWHM ~40 nm) and significant Stoke shifts (the separation between absorption and emission) (figure 8.3(C)). Through comprehensive analysis of absorption/fluorescence spectra and time-resolved fluorescence decay, we observed efficient FRET within the Pdots, resulting in a final quantum yield exceeding 30%. Leveraging single-particle fluorescence measurements and flow cytometry of Pdot-labeled cells, we demonstrated that these NIR-emitting Pdots

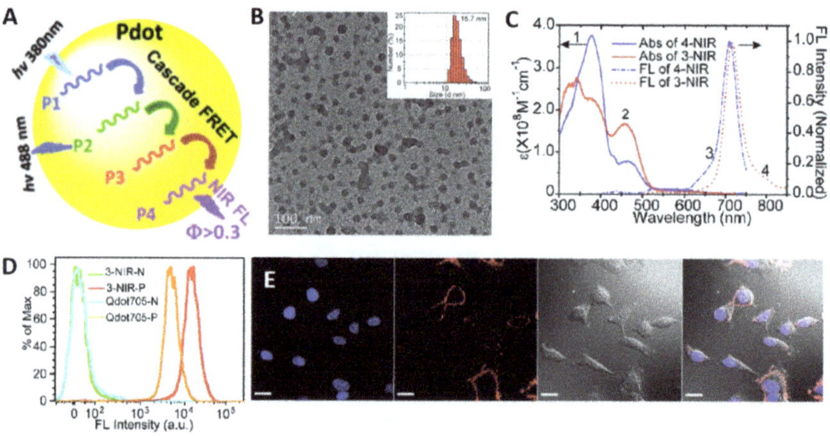

Figure 8.3. NIR-emissive Pdots. (A) Schematic of cascade FRET in a Pdot containing four polymers (P1–P4) which are FRET donors, acceptors, or energy transfer units. (B) Transmission electron microscope (TEM) image of '3-NIR' Pdots (created by blending three polymers, P1, P2, and P4). Inset: size distribution. (C) Absorbance spectra (solid lines) and fluorescence spectra (dashed lines) of '4-NIR' Pdots (prepared by blending four polymers, P1–P4) (blue, λ_{ex} = 380 nm) and 3-NIR Pdots (red, λ_{ex} = 450 nm). (D) Intensity distributions of MCF-7 breast cancer cells labeled with Qdot 705-streptavidin (negative labeling, cyan curve; positive labeling, orange curve) and 3-NIR Pdot-streptavidin (negative labeling, green curve; positive labeling, red curve), measured using flow cytometry. In negative labeling, primary biotinylated antibodies were not used. (E) Confocal fluorescence images of MCF-7 cells positively labeled with 3-NIR Pdot-streptavidin probes. Images from left to right: blue fluorescence from a nuclear stain (Hoechst 34580); deep red fluorescence from 3-NIR Pdot-streptavidin probes; DIC image; combined fluorescence image. Scale bars, 20 μm. Adapted from [15] with permission from the Royal Society of Chemistry.

showed brightness several times higher compared with Qdot 705 (figure 8.3(D)). Confocal fluorescence images of cells labeled with the NIR-emissive Pdots displayed exceptional brightness while minimizing cell autofluorescence (figure 8.3(E)). This work provided a general strategy for preparing NIR-emitting Pdots suitable for various applications, and demonstrated several advantages of Pdots: (1) exceptional light-harvesting capability leading to elevated brightness; (2) efficient cascade FRET within Pdots, enabling tunable emission with significant Stoke shifts; and (3) a hydrophobic Pdot interior which serves as an optimal host for NIR emitters, effectively overcoming solubility challenges for NIR emitters in aqueous solutions.

Pdots with time delayed emission can be used to minimize background by time-gating out autofluorescence and other sources of background noise, which have a short emission lifetime [16]. Long luminescence lifetime, also known as long excited-state lifetime, results from the decay of triplet states of emitters, and typically ranges from microseconds to milliseconds but can be even longer. Many lanthanide compounds exhibit such long luminescence lifetimes [17], and lifetimes as long as minutes or even hours can be achieved by solid materials doped with rare earth ions [18]. In contrast, autofluorescence exhibits a significantly shorter lifetime, typically on the order of nanoseconds. We designed a new type of Pdot by incorporating a europium (Eu) lanthanide complex into the host semiconducting polymer poly-vinylcarbazole (PVK) [16]. We observed that PVK served a dual role. It functioned as an ideal host matrix, effectively dispersing Eu complex molecules and minimizing self-quenching. Simultaneously, PVK proved to be an efficient light harvester and energy transfer donor, effectively stimulating the emission of the Eu complex. The resulting Eu complex/PVK Pdots exhibited an ultra-narrow red emission bandwidth of only 8 nm and an impressively high emission brightness with a quantum yield of over 30%. Notably, these Pdots displayed a remarkably long luminescence lifetime of ~500 μs. Leveraging this extended luminescence lifetime, we successfully utilized time-gated fluorescence microscopy, enabling us to distinguish the small Pdots (<20 nm diameter) from a background of larger (300 nm) bright fluorescent beads. Both the Pdots and the beads emitted in the same wavelength range, with an emission peak at 612 nm. However, the Pdots' significantly longer fluorescence lifetime (500 μs) distinctively contrasted with the much shorter fluorescence lifetime of the beads, which was approximately 3.6 ns. Images of cells labeled with these Pdots showed excellent signal-to-noise ratios and greatly reduced background fluorescence versus ungated images (figure 8.4). The signal-to-noise ratio was enhanced from 85 in the ungated image to an impressive 232 in the time-gated image.

Another approach to improve assay signal-to-noise ratio is to use Pdots with emission peaks at two different wavelengths that allow ratiometric measurements and internal calibration. For example, we designed a ratiometric metabolite biosensor for real-time point-of-care measurements of patient metabolite levels that combines an NAD(P)H-sensitive Pdot and a metabolite-specific NAD(P)H-dependent enzyme in a solution- or paper-based assay [19]. Over 100 medically relevant metabolites can be quantified by measuring the stoichiometric formation of NAD(P)H in enzyme-catalyzed reactions. This biosensor can be used to improve the diagnosis and management of a variety of diseases including diabetes,

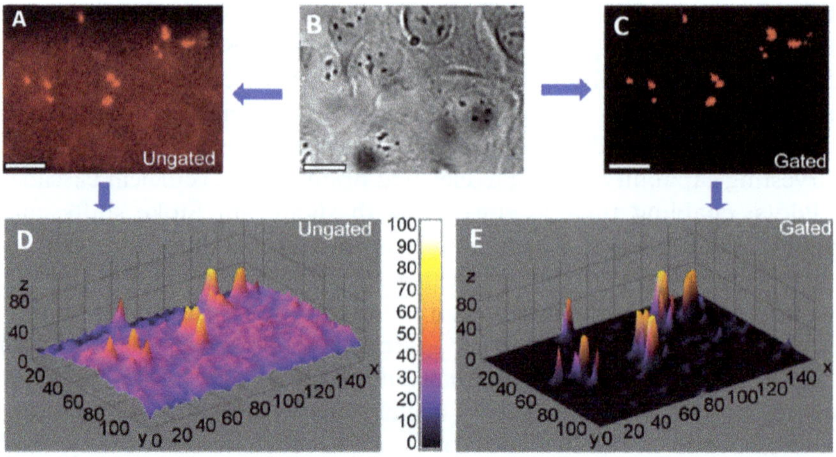

Figure 8.4. Pdots with long luminescence lifetimes for removing cell autofluorescence and increasing signal-to-noise ratio. (A–C) Ungated fluorescence image, bright field image, and time-gated fluorescence image of MCF-7 cells labeled with Eu complex/polyvinylcarbazole Pdots. Scale bar, 50 μm. (D) Three-dimensional (3D) surface plot of ungated image from panel A. (E) 3D surface plot of gated image from panel C. Reproduced from [16] John Wiley & Sons. Copyright (2009).

phenylketonuria, and nonalcoholic fatty liver disease. This biosensor works by combining NAD(P)H-sensitive Pdot with a metabolite-specific NAD(P)H-dependent enzyme, which catalyzes the oxidation of the target metabolite, resulting in NAD (P)H production. Under UV illumination, the NAD(P)H quenches the red emission of the Pdot through electron transfer while also fluorescing in the blue region (figure 8.5). Metabolite concentration is quantified ratiometrically by analyzing the ratio of blue-to-red channel emission intensities, utilizing a digital camera and RGB image processing to ensure high sensitivity and specificity. The NAD(P)H-sensitive Pdot was prepared using the luminescent conjugated polymer DPA-CN-PPV (CN-PPV is defined in the figure 8.1 caption; DPA is {2,5-bis(N,N′-diphenylamino)-1,4-phenylene}) and the amphiphilic polymer poly(styrene-co-maleic anhydride). We demonstrated accurate phenylalanine biosensing in human plasma, highlighting its application in a phenylketonuria screening test as well as biosensors for the disease-related metabolites lactate, such as glucose, glutamate, and β-hydroxybutyrate. Additionally, we have developed a paper-based version of the assay, compatible with smartphone imaging for point-of-care use. This versatile biosensor holds immense potential, as it can effectively detect any metabolite that undergoes enzymatic oxidation by NAD+ or NADP+.

8.3 Pdot preparation and bioconjugation

Pdot preparation involves inexpensive commercial materials and can be performed reproducibly and at large scale. Conjugated polymer nanoparticles can be produced in two ways: direct polymerization from low-molecular-weight monomers, or postpolymerization methods using high-molecular-weight polymers that have already been synthesized [20]. The method of direct polymerization traces its roots

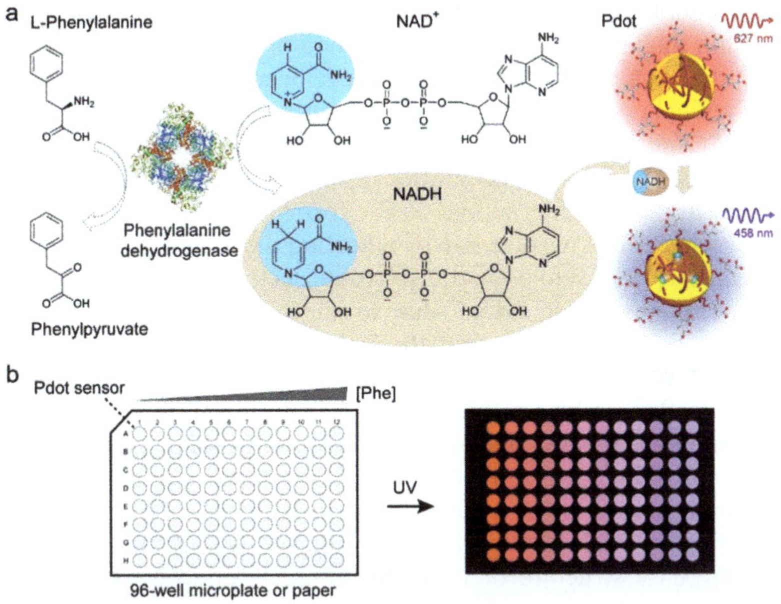

Figure 8.5. Pdot metabolite biosensor that uses ratiometric measurements and minimal equipment. (A) Phenylalanine sensing mechanism for a phenylketonuria screening test. The catalysis of L-phenylalanine by phenylalanine dehydrogenase, resulting in the formation of NADH. NADH, in turn, causes the quenching of Pdot fluorescence emission at 627 nm through electron transfer, while also emitting fluorescence at 458 nm. (B) As the phenylalanine concentration increases, fluorescence emission upon UV excitation shifts from red (Pdot emission) to blue (NADH emission). Metabolite concentration is measured ratiometrically based on the ratio of blue-to-red channel emission intensities utilizing a digital camera or plate reader, in solution- or paper-based assay formats. Reproduced from [19] John Wiley & Sons. Copyright (2021).

back to the 1980s [21, 22], and this approach was more recently extended to encompass the synthesis of fluorescent semiconducting polymers for nanoparticle preparations [23–26]. In contrast to conductive polymers, which can be synthesized through aqueous oxidative polymerization, the preparation of fluorescent semiconducting polymers typically involves coupling reactions catalyzed by transition metals. In direct polymerization, the reactions must be compatible with the dispersion solvents; in contrast, postpolymerization can draw from the vast library of commercial semiconducting polymers and does not require specialized equipment or expertise in organic and polymer synthesis. Pdots have been prepared mainly by postpolymerization approaches.

Postpolymerization methods for Pdot preparation include miniemulsion and reprecipitation methods, wherein semiconducting polymers dissolved in organic solvent usually serve as the starting solution. Water is the preferred final dispersion medium for nanoparticles synthesized for biological applications. In the miniemulsion approach, semiconducting polymer nanoparticles are formed from an emulsified droplet solution, typically using a water-immiscible solvent. On the other hand, reprecipitation methods involve precipitating the polymer by rapidly mixing the polymer solution with water, using a water-miscible solvent. The sudden reduction in solubility, coupled with hydrophobic interactions between polymer chains or segments within single chains, leads to the formation of a stable suspension of Pdots. This reprecipitation method was originally developed by Masuhara and coworkers to create polythiophene particles ranging in diameter from 40–400 nm [27], and was modified by McNeill and coworkers for forming smaller Pdots of 5–30 nm diameter [1, 4, 5, 7–11, 28–36]. The reprecipitation process involves a competition between interchain and intrachain interactions to the formation of nanoparticles, providing an opportunity to control particle size by adjusting the polymer concentration in the precursor solution [7, 8]. While the transmission electron microscope (TEM) images show a spherical shape for Pdots, due to the rigid backbone of π-conjugated polymers [37], it is worth noting that the large interfacial surface tension between the polymer and water can significantly influence the morphology in this size range. As a result, the thermodynamically favored shape for Pdots at this scale often manifests as a sphere [38, 39]. The inherent hydrophobic interactions and substantial surface tension in Pdots composed of hydrophobic semiconducting polymers often lead to internal structures with densely packed chromophores. This is evident through the highly efficient energy transfer observed within the nanoparticles [40]. Generally, the phase structure of Pdots tends to be amorphous (glassy phase); a crystalline phase can be formed partially in the glassy matrix for certain polymer species such as poly(9,9-dioctylfluorenyl-2,7-diyl) (PFO) [32].

Pdots can be customized for a wide variety of applications by varying the surface functionality used in bioconjugation reactions. Like Qdots, Pdots were initially functionalized by using encapsulation or embedding. Semiconducting polymers were encapsulated or embedded in materials such as silica [7], phospholipids [41, 42], and poly lactic-co-glycolic acid (PLGA) polymers [43, 44] to form functionalized nanoparticles. Silica encapsulation is widely used for surface functionalization of nanoparticles [45–47]. This approach has proven effective in producing Pdots with a particle size ranging from 10–20 nm, featuring a 2–3 nm thick silica shell enveloping the Pdot core. While the silica embedding method enables the creation of relatively small particle sizes, it is essential to consider the challenges associated with the silica shell in biological environments. Specifically, the silica shell is susceptible to hydrolysis, potentially compromising the stability of the Pdots within such settings. Additionally, amine-functionalized silica-shell surfaces can lead to nonspecific absorption between nanoparticles and proteins or cells. Encapsulation in phospholipids and PLGA results in nanoparticles larger than 30 nm with low per-particle fluorescence brightness. Despite the advantages of encapsulation strategies in certain aspects, they do not fully leverage the optical properties of semiconducting polymers to achieve nanoparticles with densely packed polymer fluorophores and maximize per-particle brightness.

The quest for a dependable functionalization method is crucial in creating small, bright Pdot probes that can be easily modified through bioconjugation to precisely recognize specific molecular and cellular targets. Our group developed several methods to address the challenge of Pdot functionalization for bioconjugation, described below.

One Pdot functionalization strategy that we used was based on the blending of heterogeneous polymer chains within the nanoparticle during its formation. In this approach, amphiphilic polymer carrying specific functional groups is coprecipitated alongside semiconducting polymer during Pdot formation, resulting in a functionalized surface. Subsequent conjugation covalently attached biomolecules which recognize cell targets via antigen–antibody or biotin–streptavidin interactions. This effective functionalization and bioconjugation strategy can be universally applied to any fluorescent semiconducting polymer. An excellent illustration of its potential was demonstrated by using the amphiphilic comb-like polymer PS-PEG-COOH to successfully functionalize PFBT Pdots (figure 8.6(A)) [11]. Functionalized Pdots were prepared using a constant PFBT concentration and PS-PEG-COOH/PFBT fractions of 0%–20% by weight. The functionalized PFBT Pdots were ~15 nm in diameter based on TEM (figure 8.6(B)) and dynamic light scattering (DLS) (figure 8.6(C)). This functionalization strategy produced nanoparticle probes with a much higher fluorophore density than that of other probes: over 80% of the Pdots by weight consisted of fluorescent polymer, whereas only a few percent by weight of the Qdots consisted of fluorescent polymer due to the presence of a thick encapsulation shell, and the effective fluorophore density of dye-loaded spheres was reduced by self-quenching of the dyes.

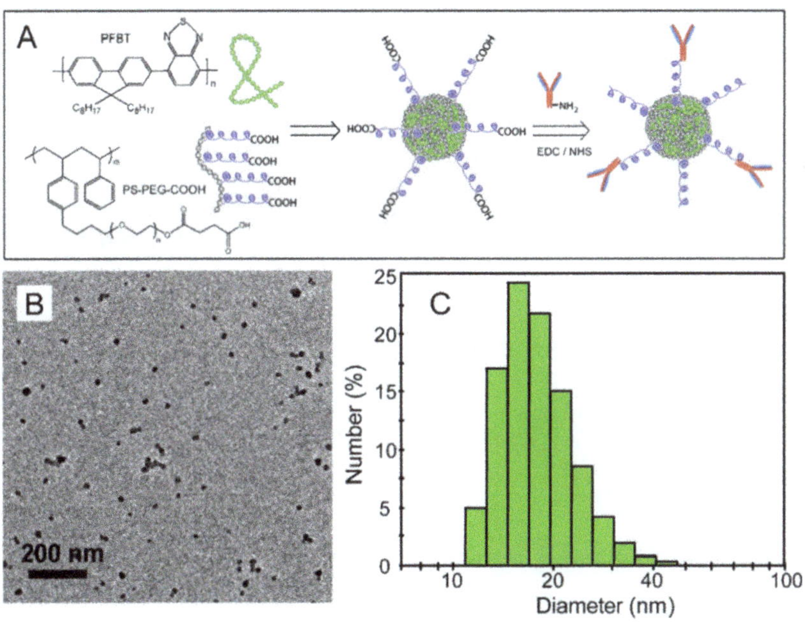

Figure 8.6. Functionalized Pdots for bioconjugation. (A) Surface functionalization of PFBT Pdots with reactive carboxy groups by coprecipitation of PFBT with a small amount of the amphiphilic polymer PS-PEG-COOH, followed by bioconjugation to an antibody primary amine via an N-Hydroxysuccinimide (NHS) ester reaction. The COOH-PFBT Pdots were ~15 nm in diameter based on TEM (B) and DLS (C). Adapted with permission from [11]. Copyright (2010) American Chemical Society.

Another method we developed for producing 'clickable' Pdots for bioconjugation involved coprecipitating PFBT with a small quantity of poly(styrene-co-maleic anhydride) (figure 8.7(A)) [4]. After introducing carboxy-functional groups to the Pdots, we further conducted reactions with amine-containing molecules, including amino azides or alkynes. Through this process, we successfully engineered Pdots suitable for bioorthogonal labeling using click chemistry. The modified and unmodified Pdots had a similar average diameter of ~15 nm, but exhibited shifted gel electrophoresis migration behavior because of the reduced charge on the modified Pdots compared to unmodified carboxy-Pdots (figure 8.7(B)). A fluorescence assay was conducted using alkyne-functionalized Alexa Fluor 594. The results demonstrated the high reactivity of the azide-functionalized Pdots towards the terminal alkyne group through a copper(I)-catalyzed click reaction (figure 8.7(C)).

In the above methods, a small amount of functionalized polymer was incorporated into a Pdot primarily through noncovalent hydrophobic interactions. The dependence on hydrophobic association and the absence of covalent attachment can

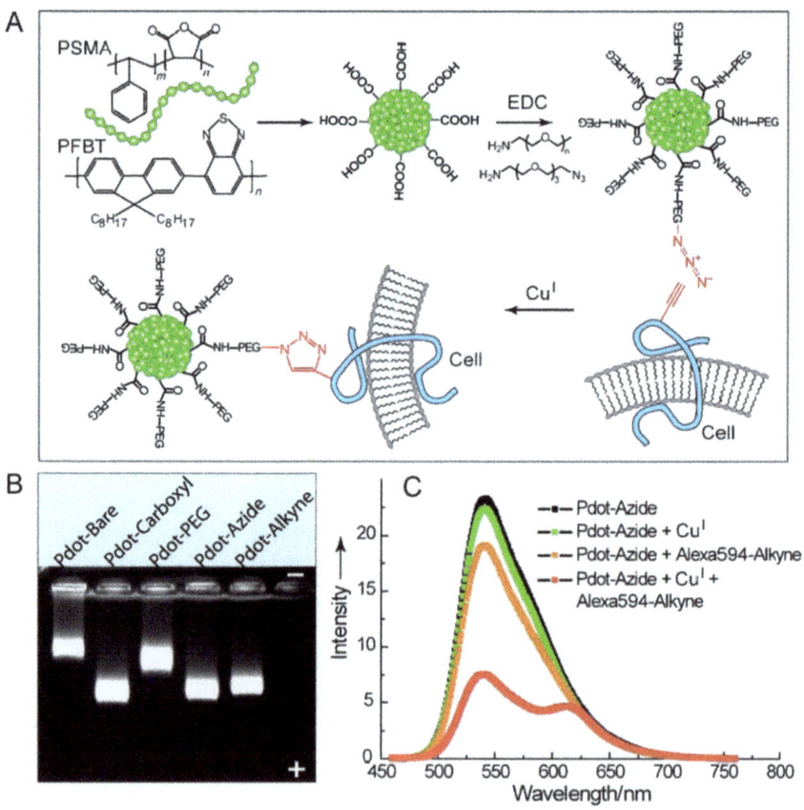

Figure 8.7. Clickable Pdots for bioorthogonal labeling. (A) Pdot functionalization and bioorthogonal labeling through click chemistry. (B) Gel electrophoresis of Pdots with different surface functional groups. (C) A fluorescent assay using alkyne-Alexa Fluor 594 to confirm the presence of azides on the Pdot surface. Adapted with permission from [4]. Copyright (2010) American Chemical Society.

pose a drawback because the functionalized polymer may dissociate from the nanoparticles due to swelling, reducing bioconjugation labeling efficiency. In addition, it is difficult to precisely control the density of functional groups on the Pdot surface when using these methods. To overcome these drawbacks, our group developed an approach called 'direct functionalization' that involves covalently attaching functional groups during the initial polymer synthesis [5]. Pdots derived from these polymers possess inherent functional groups that are readily available for direct bioconjugation, eliminating the necessity for further surface modification. Direct functionalization involves different considerations than the above methods and sometimes has opposite requirements. The degree of hydrophilic functionalization is an important determinant of both the stability and fluorescence performance of the Pdots, as we discovered through our research. To ensure optimal performance for biological applications, it is essential that the density of hydrophilic groups be carefully controlled at low levels. To delve further into the impact of hydrophilic side chains on Pdot stability and performance, we synthesized a series of PFBT polymers featuring side-chain carboxylic acid groups at molar fractions of approximately 2%, 14%, and 50%. These distinct variants were named PFBT-C2, PFBT-C14, and PFBT-C50, respectively (figure 8.8). As the density of hydrophilic functional groups increased from 2% to 50%, the fluorescence quantum yield experienced a decline from 30% to 17%. Single-particle fluorescence brightness comparisons brought to light that the actual fluorophore packing density in PFBT-C50 Pdots was lower than that observed in PFBT-C2 and PFBT-C14 Pdots. This observation suggests that the presence of hydrophilic side chains exterted an influence on the internal structures and compactness of the Pdots. To investgate and confirm this possibility, a dye-doping and leaching method was developed to examine the compactness of fluorophore packing and the association strength between polymer molecules within PFBT Pdots with varying densities of hydrophilic side chains. Through this method, we observed that dye leaching and PFBT dissociation in Pdots were notably

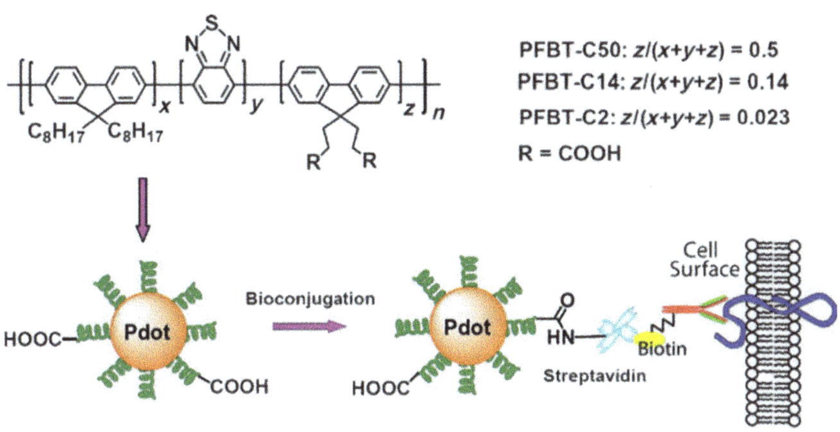

Figure 8.8. Directly functionalization of Pdots with various side-chain carboxy groups. Bioconjugation was performed on the PFBT-C2 Pdots and the resulting Pdot-bioconjugates were specific for cellular labeling. Reprinted with permission from [5]. Copyright (2012) American Chemical Society.

influenced by the density of hydrophilic side chains. For instance, approximately 10% of PFBT and 50% of dye molecules leached from PFBT-C50 Pdots. In contrast, no PFBT and only 6% of dye leached from PFBT-C2 Pdots. These findings, combined with FRET analysis, yielded significant insights into the internal structures of the Pdots. Specifically, PFBT-C2 Pdots displayed robust chain–chain associations and a densely packed internal structure, while PFBT-C50 Pdots, characterized by a high density of hydrophilic side chains, exhibited a different behavior. Flow cytometry experiments further revealed that PFBT-C50 Pdots demonstrated increased nonspecific cellular labeling compared to PFBT-C2 Pdots. These results collectively underscore the critical role played by low-density functional groups in direct functionalization of Pdots. We carried out bioconjugation with the PFBT-C2 dots, and the resulting Pdot bioconjugates exhibited specific cell targeting. This confirms that the low-density functionalization approach is important for successful bioconjugation and holds potential for various biological applications.

8.4 Applications of Pdots to oncology

Because of their high brightness and compatibility with bioconjugation reactions, Pdots can be used to improve the signal-to-noise ratio and reduce the instrumentation requirements of a wide variety of fluorescence-based diagnostic assays, including protein, nucleic acid, and cell assays. For example, many protein biomarkers are used in cancer screening, and we demonstrated highly sensitive protein detection on a paper substrate [48].

Because of their tunable optical properties, Pdots can be developed with bright NIR emissions and long Stokes shifts as described in section II to better distinguish specific cell types such as cancer cells from tissue autofluorescence *in vivo*. We applied NIR-emitting Pdots to 'tumor painting' *in vivo* [49]. Chlorotoxin (CTX), a 36-amino-acid peptide with strong affinity towards tumors of neuroectodermal origin, thus was chosen as the tumor-targeting ligand. To achieve CTX conjugation, we functionalized a polymer-blend Pdot, comprising PFBT and PF-DBT5 with the amphiphilic polymer poly(styrene-co-maleic anhydride) to introduce surface carboxyl groups. This modification facilitated the efficient conjugation of CTX using standard carbodiimide chemistry (figure 8.9(A)). To enhance the performance of Pdots *in vivo*, polyethylene glycol was conjugated to reduce protein adsorption, minimize immune recognition, and extend the nanoparticle's serum half-life. These CTX-conjugated Pdots demonstrated specificity in targeting malignant brain tumors by biophotonic imaging (figure 8.9(B)), biodistribution, and histological analysis. Another potential approach to tumor painting with Pdots is to use pH-sensitive Pdots to detect the low pH of tumor environments. We have developed two types of pH-sensitive Pdots [50, 51].

Photodynamic therapy (PDT) stands as a highly promising and minimally invasive approach for treating malignant cancer that involves light irradiation, photosensitizer, and molecular oxygen [52]. When the photosensitizer is excited, it catalyzes the conversion of oxygen in its ground state to singlet oxygen, a highly

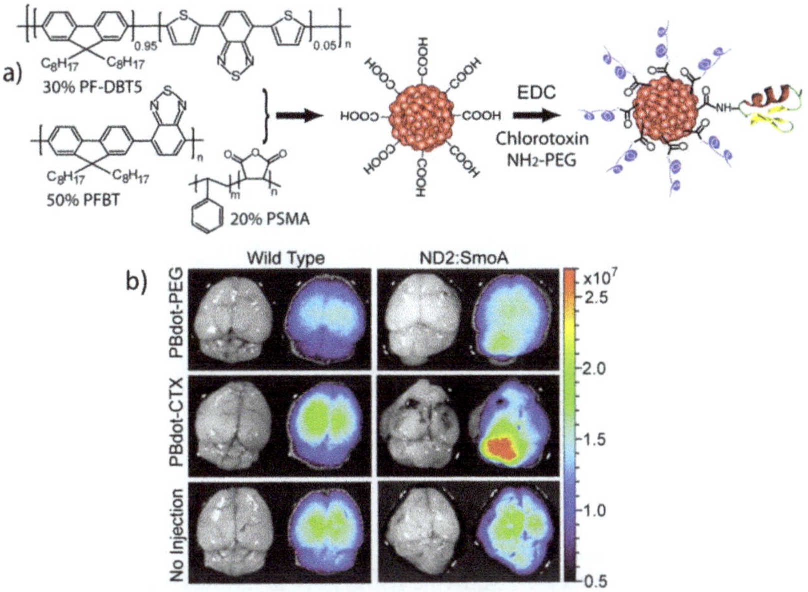

Figure 8.9. Application of Pdots to tumor painting. (A) Carboxy-functionalized, NIR-emissive Pdots were synthesized by using a mixture of PF-DBT5, PFBT, and PSMA [poly(styrene-co-maleic anhydride)] polymers. The tumor-targeting ligand CTX was conjugated to the Pdots by standard carbodiimide chemistry. (B) Fluorescence imaging of healthy brains in wild-type mice (left) and medulloblastoma tumors in a transgenic (ND2:SmoA1) mouse (right). Each mouse received a 50 μl injection of a 1 μM solution, with options of either nontargeting PBdot-PEG (top) or targeting PBdot-CTX (middle), administered through the tail vein. Control mice did not receive any injection (bottom). Adapted from [49] John Wiley & Sons. Copyright (2011).

reactive species that attacks the targeted cancer cells and tissues exposed to light. This process minimizes damage to the surrounding healthy tissue, making PDT an effective therapeutic option in the fight against cancer. The photosensitizer is a crucial component of PDT, and popular choices include porphyrin, chlorin, and their derivatives. Pdots is a good platform for developing PDT photosensitizers due to their advantageous properties, such as large extinction coefficients [53–57]. Huang and coworkers reported a series of Pdots created by conjugating the phosphorescent Ir(III) complex with polyfluorene units in the conjugated polymer chain. The Pdots demonstated high photostability and biocompatibility. Through efficient energy transfer from polyfluorene to the Ir(III) complex, these Pdots showed singlet oxygen generation upon light irradiation, leading to apoptosis of cancer cells [53]. The same research group engineered a Pdot-based photosensitizer by combining Pt(II) porphyrin as a hydrophobic, oxygen-responsive phosphorescent group with a hydrophilic, polyfluorene-based hyperbranched conjugated polyelectrolyte. The self-assembly of these components resulted in the formation of nanoparticle structures. The Pdots containing Pt(II) complexes displayed ratiometric luminescence, as they exhibited both fluorescence from polyfluorene and phosphorescence from Pt(II) complexes, providing an effective means of oxygen sensing. O_2 levels were quantified through the ratiometric emission intensity of phosphorescence and fluorescence, as well as the measured phosphorescence lifetime. Under hypoxic conditions, cells treated with Pdots exhibited longer fluorescence lifetimes, presenting an opportunity for improved oxygen sensing. Time-resolved luminescence images showed an enhanced signal-to-noise ratio achieved by gating off the short-lived background fluorescence. The Pdots displayed effective PDT activity, validated through a combination of flow cytometry, cell viability assays, and imaging of PDT-induced cell death. The Wu group showed a significant enhancement in singlet oxygen generation within a Pdot-based photosensitizer designed for PDT [54, 55]. The incorporation of tetraphenylporphyrin (TPP) monomers into the PFBT π-conjugated backbone led to the formation of PFBT-TPP Pdots. These Pdots have a high $^1O_2$ generation quantum yield of 35%, efficiently elimating cancer cells *in vitro*, and displaying significant therapeutic activity in tumor-bearing mice *in vivo*, with a noteworthy inhibition of tumor growth.

Glossaries

Abbreviations	Full Meaning
CN-PPV	poly[2-methoxy-5-(2-ethylhexyloxy)-1,4-(1-cyanovinylene-1,4-phenylene)]
CTX	chlorotoxin
DLS	dynamic light scattering
Eu	europium
ϕ_B	quantum yield
FRET	Förster resonance energy transfer
LED	light-emitting diode
LMICs	low - and middle - income countries
MEH-PPV	poly[2-methoxy-5-(2-ethylhexyloxy)-1,4-phenylenevinylene]
NAD+	nicotinamide adenine dinucleotide
NADH	nicotinamide adenine dinucleotide hydrogen
NIR	near-infrared
PCR	polymerase chain reaction
PDHF	poly(9,9-dihexylfluorenyl-2,7-diyl)
Pdots	semiconducting polymer dots
PDT	photodynamic therapy
PF-DBT5	poly(9,9-dioctylfluorene)-co-(4,7-di-2-thienyl-2,1,3-benzothiadiazole)
PFBT	poly[(9,9-dioctylfluorenyl-2,7-diyl)-co-(1,4-benzo-{2,1,3}-thiadiazole)
PFO	Poly(9,9-dioctylfluorenyl-2,7-diyl)
PFPV	poly[{9,9-dioctyl-2,7-divinylene-fluorenylene}-alt-co-{2-methoxy-5-(2-ethylhexyloxy)-1,4-phenylene}]
PLGA	poly lactic-co-glycolic acid
PPE	poly(phenylene ethynylene)
PVK	polyvinylcarbazole
Qdots	quantum dots
TEM	transmission electron microscopes
THF	tetrahydrofuran
TPP	tetraphenylporphyrin
UV	ultraviolet

References

[1] Jin Y, Ye F, Wu C, Chan Y H and Chiu D T 2012 *Chem. Comm.* **48** 3161–3

[2] Ye F, White C C, Jin Y, Hu X, Hayden S, Gao X, Kavanagh T J and Chiu D T 2015 *Nanoscale* **7** 10085–93

[3] Sun W, Ye F, Gallina M E, Yu J, Wu C and Chiu D T 2013 *Anal. Chem.* **85** 4316–20

[4] Wu C, Jin Y, Schneider T, Burnham D and Chiu D T 2010 *Angew. Chem. Int. Ed.* **49** 9436–40

[5] Zhang X, Yu J, Wu C, Jin Y, Ye F and Chiu D T 2012 *ACS Nano* **6** 5429

[6] Hide F, Diaz-Garcia M A, Schwartz B J and Heeger A J 1997 *Acc. Chem. Res.* **30** 430

[7] Wu C, Szymanski C and McNeill J 2006 *Langmuir* **22** 2956

[8] Wu C, Bull B, Szymanski C, Christensen K and McNeill J 2008 *ACS Nano* **2** 2415–23

[9] Wu C, Hansen S, Hou Q, Yu J, Zeigler M, Jin Y, Burnham D, McNeill J, Olson J and Chiu D T 2011 *Angew. Chem.* **123** 3492

[10] Ye F, Wu C, Jin Y, Wang M, Chan Y, Yu J, Sun W, Hayden S and Chiu D T 2012 *Chem. Commun.* **48** 1778

[11] Wu C, Schneider T, Zeigler M, Yu J, Schiro P, Burnham D, McNeill J D and Chiu D T 2010 *J. Am. Chem. Soc.* **132** 15410

[12] Eggeling C, Widengren J, Rigler R and Seidel C A M 1998 *Anal. Chem.* **70** 2651

[13] van Sark W G J H M, Frederix P L T M, Van den Heuvel D J, Gerritsen H C, Bol A A, van Lingen J N J, Donega C D and Meijerink A 2001 *J. Phys. Chem.* B **105** 8281

[14] Jin Y, Ye F, Zeigler M, Wu C and Chiu D T 2011 *ACS Nano* **5** 1468–75

[15] Zhang X, Yu J, Rong Y, Ye F, Chiu D T and Uvdal K 2013 *Chem. Sci.* **4** 2143–51

[16] Sun W, Yu J, Deng R, Rong Y, Fujimoto B, Zhang H and Chiu D T 2013 *Angew. Chem. Int. Ed.* **52** 11294–7

[17] Binnemans K 2009 *Chem. Rev.* **109** 4283–374

[18] Aitasalo T, Deren P, Holsa J, Jungner H, Krupa J C, Lastusaari M, Legendziewicz J, Niittykoski J and Strek W 2003 *J. Solid State Chem.* **171** 114–22

[19] Chen H, Yu J, Zhang J, Sun K, Ding Z, Jiang Y, Hu Q, Wu C and Chiu D T 2021 *Angew. Chem. Int. Ed.* **60** 19331–6

[20] Pecher J and Mecking S 2010 *Chem. Rev.* **110** 6260

[21] Edwards J, Fisher R and Vincent B 1983 *Makromol. Chem. Rapid Commun.* **4** 393

[22] Armes S P, Miller J F and Vincent B 1987 *J. Colloid Interface Sci.* **118** 410

[23] Baier M C, Huber J and Mecking S 2009 *J. Am. Chem. Soc.* **131** 14267–73

[24] Pecher J and Mecking S 2007 *Macromolecules* **40** 7733

[25] Pecher J, Huber J, Winterhalder M, Zumbusch A and Mecking S 2010 *Biomacromolecules* **11** 2776

[26] Hittinger E, Kokil A and Weder C 2004 *Angew. Chem. Int. Ed.* **116** 1844

[27] Kurokawa N, Yoshikawa H, Hirota N, Hyodo K and Masuhara H 2004 *Chem. Phys. Chem.* **5** 1609

[28] Szymanski C, Wu C, Hooper J, Salazar M A, Perdomo A, Dukes A and McNeill J D 2005 *J. Phys. Chem.* **109** 8543

[29] Wu C, Peng H, Jiang Y and McNeill J 2006 *J. Phys. Chem.* B **110** 14148

[30] Wu C, Szymanski C, Cain Z and McNeill J 2007 *J. Am. Chem. Soc.* **129** 12904

[31] Wu C, Zheng Y, Szymanski C and McNeill J 2008 *J. Phys. Chem.* C **112** 1772

[32] Wu C and McNeill J 2008 *Langmuir* **24** 5855

[33] Wu C, Bull B, Szymanski C, Christensen K and McNeill J 2009 *Angew. Chem. Int. Ed.* **121** 2779–83
[34] Yu J, Wu C, Sahu S, Fernando L, Szymanski C and McNeill J 2009 *J. Am. Chem. Soc.* **131** 18410–14
[35] Yu J, Wu C, Tian Z and McNeill J 2012 *Nano Lett.* **12** 1300
[36] Yu J, Wu C, Zhang X, Ye F, Gallina M E, Rong Y, Wu Y, Sun W, Chan Y-H and Chiu D T 2012 *Adv. Mater.* **24** 3498
[37] Schwartz B J 2003 *Annu. Rev. Phys. Chem.* **54** 141
[38] Chandler D 2005 *Nature* **437** 640
[39] ten Wolde P R and Chandler D 2002 *Proc. Natl. Acad. Sci. USA* **99** 6539
[40] Tian Z, Yu J, Wu C, Szymanski C and McNeill J 2010 *Nanoscale* **2** 1999
[41] Howes P, Green M, Levitt J, Suhling K and Hughes M 2010 *J. Am. Chem. Soc.* **132** 3989–96
[42] Howes P, Green M, Bowers A, Parker D, Varma G, Kallumadil M, Hughes M, Warley A, Brain A and Botnar R 2010 *J. Am. Chem. Soc.* **132** 9833
[43] Li K, Pan J, Feng S S, Wu A W, Pu K Y, Liu Y T and Liu B 2009 *Adv. Funct. Mater.* **19** 3535
[44] Li K, Liu Y T, Pu K Y, Feng S S, Zhan R Y and Liu B 2011 *Adv. Funct. Mater.* **21** 287
[45] Bruchez M, Moronne M, Gin P, Weiss S and Alivisatos A P 1998 *Science* **281** 2013
[46] Ow H, Larson D R, Srivastava M, Baird B A, Webb W W and Wiesner U 2005 *Nano Lett.* **5** 113
[47] Wang L, Yang C Y and Tan W H 2005 *Nano Lett.* **5** 37
[48] Ye F, Smith P B, Wu C and Chiu D T 2013 *Macromol. Rapid Commun.* **34** 785–90
[49] Wu C, Hansen S J, Hou Q, Yu J, Zeigler M, Jin Y, Burnham D R, McNeill J D, Olson J M and Chiu D T 2011 *Angew. Chem. Int. Ed.* **50** 3430–4
[50] Chen L, Wu L, Yu J, Kuo C T, Jian T, Wu I, Rong Y and Chiu D T 2017 *Chem. Sci.* **8** 7236–45
[51] Chan Y H, Wu C, Ye F, Jin Y, Smith P B and Chiu D T 2011 *Anal. Chem.* **83** 1448–55
[52] Dolmans D E J G J, Fukumura D and Jain R K 2003 *Nat. Rev. Cancer* **3** 380
[53] Shi H, Ma X, Zhao Q, Liu B, Qu Q, An Z, Zhao Y and Huang W 2014 *Adv. Funct. Mater.* **24** 4823–30
[54] Chang K, Tang Y, Fang X, Yin S, Xu H and Wu C 2016 *Biomacromolecules* **17** 2128–36
[55] Li S, Chang K, Sun K, Tang Y, Cui N, Wang Y, Qin W, Xu H and Wu C 2016 *ACS Appl. Mater. Interfaces* **8** 3624–34
[56] Zhang D, Wu M, Zeng Y, Liao N, Cai Z, Liu G, Liu X and Liu J 2016 *J. Mater. Chem.* B **4** 589–9
[57] Zhou X *et al* 2016 *Adv. Sci. (Weinheim, Ger.)* **3** 1500155

Chapter 9

Humanitarian engineering education

Claire Dempsey, Derek Brown and Adam Shulman

9.1 Introduction

Creating a fit-for-purpose medical workforce in low- and middle-income countries (LMICs) is a challenging task on many levels. Much consideration is often directed toward equipment and technological measures but the importance of educating both this generation and the next generation of medical experts cannot be overlooked [1]. Health education needs to be sensitive to local population needs, both medically and culturally. Such training should inherently consider the following components: goals/objectives, program certification/recognition, curriculum outline/content, methods of assessment, funding, logistics, educational resources, and outcomes (including analysis of educational and clinical patient-related outcomes) as well as acknowledge the challenges in implementation and sustainability.

There are several methods of training that have been used in LMICs across a range of medical specialties, each with varying levels of success. Some models consider the health workforce at an undergraduate/postgraduate level while others focus on expanding the professional development of existing clinical staff. In each of these, the creation of local LMIC medical experts has relied heavily on partnerships between higher- and lower-resource settings. In most professional modalities there is a diverse range of educational offerings spanning from short courses to full residency programs.

9.2 Methods of education and training

The best methodology for providing a meaningful learning experience for LMIC participants can vary widely based on the particular knowledge and skills educators are trying to teach. Didactic teaching remains the instructional mainstay of many educational programs; however, this may not be the most efficient or effective use of the educator's or student's time if there is not a community-engaged and relevant curriculum to teach. Identifying fundamental competencies that are required to meet the evolving needs of the LMIC population is the first key step. Task-specific

doi:10.1088/978-0-7503-3751-9ch9

training may be of greater benefit to LMIC departments than generalized teaching curriculums. This type of training will allow clinics to implement quality improvement measures in a meaningful and supervised way.

A teaching strategy that is now becoming the norm is e-learning. E-learning is a tool that can be used in both high-income and resource-constrained countries, particularly when delivered in an open access environment. Studies have demonstrated the positive effects of e-learning or blended-learning courses compared to more traditional didactic teaching in the acquisition and retention of knowledge [2, 3]. E-learning is also easily accessible by healthcare workers in remote areas and can allow them to continue developing skills through distance education. However, not all competencies can be mastered without some form of personalized interaction and feedback from educators. In these cases it may be appropriate to incorporate blended or even traditional teaching strategies [4].

Simulation and problem-based learning are useful tools to help accelerate the acquisition of skills by participants [5]. This type of training is particularly useful for practicing procedures that participants would otherwise not be exposed to in the clinic, or in preparation for implementation of new technology or techniques. They allow for a variety of situations and are specially designed for the development of skills that can only be learned through participation. Integration of these training methods requires experienced educators to teach, supervise, and evaluate. It may also require access to anonymized patient information or special software platforms.

While online learning is an efficient and effective use of both educator and trainee time, there is no substitute for on-site training. Ideally, this training should be conducted at the trainee's institution to allow the educator the opportunity to experience the current treatment workflow and understand any resource implications, as it may be necessary to modify international best practice recommendations to meet the challenges of the local environment. It is recognized that time and financial resource restrictions may result in on-site training not always being available.

Traditionally, education of different professional streams has been conducted in silos, with each group developing their own set of competencies, creating a culture of ownership over a specific piece of work. However, multidisciplinary teamwork is required as part of effective care delivery and, as such, interprofessional education and training may be appropriate in some instances [6]. The experience of learning together can enhance collaboration and teamwork, and ultimately improve patient treatment quality [7]. Interprofessional education can ensure that all team members understand the role that others play in the patient treatment journey.

There are currently multiple methods for educating and training the clinical workforce in LMICs, beyond that which is provided by local educators. Each scheme has positive and negative aspects that need to be taken into consideration when contemplating the most appropriate way to support the development of global education programs.

Some methods of training include:

- *LMIC students completing degrees (undergraduate and/or postgraduate) in developed countries using funding support models*

  - Positives
    - Assurance that the quality of education meets professional standards
    - Education with exposure to modern equipment and techniques
    - Creation of a wide professional network from participants and educators
    - Participants receive an internationally recognized qualification on completion.
  - Negatives
    - Participants are removed from local support networks (family/friends)
    - Risk of culture shock living in a different country
    - Risk that participants will not return to their local country once qualification is received as the likelihood of employment opportunities elsewhere increases with higher qualifications
    - Can be costly to fund education, housing, and living expenses over several years for each participant, particularly living in developed countries/cities
    - Teaching content may not match the technology, treatment techniques, or patient cohorts of the LMIC.
- *Hosting LMIC staff in clinics in developed countries*
  - Positives
    - Training clinical staff in current best practice in a highly functioning department
    - Helpful when implementing new treatment techniques or integration of new equipment in the LMIC clinic to observe established work processes
    - Building a lasting connection with another clinical department for long-term support.
  - Negatives
    - Difficulties finding a teaching department with the same (or very similar) equipment profile to make the experience meaningful to the participants
    - LMIC clinical staff are removed from their department, increasing the workload for other staff
    - Can be costly to provide housing and living expenses for each participant
    - Risk that clinical exposure does not match the technology, treatment techniques, or patient cohorts of the LMIC.
- *Hosting conferences in LMICs, by LMICs*
  - Positives
    - Ownership of teaching and learning by both conveners and delegates
    - Presentations that are relevant to LMIC equipment, techniques, and patient profiles

        - Low financial outlay for both conveners and delegates
        - Networking opportunity for departments within a local region.
    - Negatives
        - Time commitment and enthusiasm required by conveners for conference organization
        - Financial burden on conveners
        - Possible lack of advanced clinical processes in the presentations
        - Possible lack of facilities to host conferences in LMIC department/region.
- *Hosting conferences in LMICs, by developed countries*
    - Positives
        - Teaching concepts directly related to the LMIC setting
        - Experts can have meaningful conversations and demonstrations beyond the structured course material, while on-site for an extended period of time
        - The opportunity exists to provide hands-on practicum under the direct supervision of travelling experts.
    - Negatives
        - Very expensive for one week of work
        - A thoughtful approach is still necessary to understand WHAT needs to be taught and HOW to teach it most effectively. Just as important is a thoughtful approach to what DOES NOT need to be taught. Otherwise, just having experts on-site does no good.
- *Supporting attendance of LMIC staff at international conferences/workshops*
    - Positives
        - Exposure to newest techniques and technology in the field
        - Networking with professionals from developing countries
        - Increasing interest in research and development.
    - Negatives
        - Financial burden must be supported through sponsorship
        - Opportunity limited to only a few participants
        - Conference presentations may not be relevant for the LMIC department conditions.
- *Brief (1–4 week) missions to LMIC centers*
    - Positives
        - In-person teaching/training can be more impactful than online or other distance-based forms of training
        - Can build positive relationships
        - Can assess and troubleshoot real working environments.
    - Negatives
        - Most effective as a part of a larger, distance-based training program
        - High costs associated with travel, for only a short interval of training

        - It can sometimes take a week or more for a visiting expert to understand the new environment before meaningful teaching can take place (this can be all or part of the visit).
- *Long-term (e.g. 1 year) training missions by individuals or teams to LMICs*
    - Positives
        - Expert, hands-on clinical training in the environment of the LMIC with the equipment and patient cohort of the LMIC
        - Improving clinical practice under the safe supervision of the visiting expert
        - LMIC staff have time to thoroughly learn and refine their skills
        - A continued relationship between the education and LMIC beyond the on-site training period is almost guaranteed
            - *'I have friends for a lifetime. More important than the change I was able to make at Sweden Ghana Medical Centre in Accra, Ghana, was the inspiration I could bring to my colleagues there—that they have the power to make a significant difference in the world. Not only were they able to perform exceptional medical physics within their department, but they became leaders in the region and on the continent. They have become my lifelong colleagues in the fight to improve radiation oncology across Africa and are confident and passionate in that endeavor.'* Adam Shulman, visiting medical physicist at Sweden Ghana Medical Centre from 2014–2015.
    - Negatives
        - Challenging to engage the services of experienced clinical staff willing to commit long-term to training in an LMIC
        - Expatriate recruitment can be expensive because of the total packages offered to the experienced educator (salary, housing, medical insurance, taxes, etc.).
        - May not address the issues associated with running an efficient center.
- *Mentorship partnerships between LMICs and clinics in developed countries*
    - Positives
        - Ability to significantly improve the quality of a single radiation therapy department
        - Oversight of all clinical practices across all professional groups
        - Not time dependent, can slowly develop skills in the LMIC department and maintain them
        - Peer review, document and procedure sharing more likely.
    - Negatives
        - Large resource burden for the developed country institution
        - Lack of continued enthusiasm for supporting the training program from the developed country institution as clinical workload takes precedence

        - Lack of oversight and responsibility can decrease training productivity
        - Human resources are high relative to the impact, as there may be multiple trainers teaching only a few LMIC staff.
- *Formal online training programs*
    - Positives
        - Experts can teach large LMIC groups in geographically distant locations simultaneously. Requires only an internet connection
            - *'Building RCC training programs has been a team approach. It's taken the input of hundreds of clinics, of multiple types of clinic staff, across multiple world regions, and some of the most passionate and experienced educators globally, and combined it into its education and training programs. What is education? Education is what survives when what has been learned has been forgotten. It doesn't depend on memorization. It is a lasting impact, and that is what we strive for. What is training? Training is developing the skills and knowledge that relate to specific useful competencies. Lots of information is interesting, but when it comes to patient care and limited resources, we have to focus our attention on what is actually helpful. By labeling our training programs, we create the mindset of improvement toward specific goals.'* Ben Li, Radiation Oncologist Resident, Founder and President of Rayos Contra Cancer
        - LMICs and experts can share clinical environments at both the trainer and trainee sites
        - LMICs can witness what a well-run clinical environment looks like while expert can troubleshoot specific issues at the LMIC sites
        - Efficient use of funds to train multiple LMICs at once.
    - Negatives
        - Internet connections and electricity may not be stable enough in some LMIC environments to provide adequate means of communication
        - Forming relationships and maintaining interest in the project can be challenging in an online environment
        - Teaching content may not be as relevant to LMICs with limited resources.
- *Online resources without training guidance*
    - Positives
        - Easy access to information in a low-income setting
        - Can reach a large number of learners from disparate geographical regions
        - Very inexpensive
        - Can combine many of the benefits of a 'formal online training program' without the need for a large number of instructors.
    - Negatives
        - Difficult to engage the audience and keep interest

- Not all concepts can be mastered without some form of personal interaction
- Different people learn better in different ways. It is difficult to create a 'one size fits all' platform.

9.3 Engagement and sustainability

As with any training endeavor, it is important to be thoughtful and creative about teaching methods and content. Program developers must be realistic about obstacles that may arise (e.g. equipment access), and propose methods for overcoming obstacles (e.g. matching the curriculum to the equipment profile in the clinic). Identifying the minimum competence required to be considered functional, safe, and problem solvers in the clinic is important in setting training standards [8]. At this point, the training program can be designed, with efficiency as its focus.

Ensuring high participant motivation is critical in a successful training program [9]. From the outset it must be clear to participants what they can gain from the program that they cannot gain from existing programs. Program developers must consider how much time participants have to undertake the program, any local deadlines that need to be met (e.g. installation of new equipment or commencement of new treatment techniques), and what could be used to hold their attention and increase engagement. Being able to modify a program to suit the needs of a particular audience is of great importance to ensure success.

The global need for training far exceeds the resources available for training, and so any resources must be used wisely, with a focus on ensuring sustainability. Not only should the probability of long-term clinical benefits for the LMIC undertaking training be considered, it is also important to consider how initial education and funding could generate additional training opportunities within the region. By educating health professionals so that they are *globally competent and locally relevant* they will be able to serve their local communities in an effective manner, thus providing the most gain for any financial expenditure. It is believed that the best model of sustainable training is to use training resources to establish and nurture regional training centers, thus creating a local and self-sustaining LMIC educational framework. This has the added benefit of providing LMICs a sense of ownership over their training.

Education and training in LMICs are multidimensional processes that involve not only increasing the knowledge, skills, and competencies that are most relevant to the needs of the LMIC population but also require development of a workforce of educators and trainers to provide effective education methods and access to adequate infrastructure and learning tools.

9.4 Infrastructure requirements

9.4.1 Human

Most LMIC training and education programs are built on the backbone of volunteers looking to have a meaningful impact in their clinical profession. The first key to successful and efficient use of human resources is to maximize volunteers' unique strengths and abilities, utilizing people in roles that they perform best. For example: one volunteer may not be a good teacher but has valuable administration and project management skills, and another volunteer may have expertise in a small area of a wider clinical service. When designing a training program, both administrative and educational tasks should be considered. It may be easier to find less experienced staff (e.g. medical students/residents) for relatively routine administration tasks than to use the resources of expert clinical staff. Regardless of who does each task, all volunteers should be provided with clear directions and expectations of the level of time required to complete the task so that volunteers can understand the commitment they are making to the program. This is important in ensuring volunteer retention and engagement into the future.

Other human resource requirements to be considered include:

- Clinical and nonclinical staffing needs
    - Delivering high-quality patient treatment requires a highly functioning, multidisciplinary team that includes both clinical and nonclinical employees. It is important to remember that any training program designed to improve quality of care in a specific department should, where possible, include both clinical and nonclinical staff.
- Enough staff in the LMIC department to perform clinical duties, as well as to learn
    - Learning requires time and effort and highly busy staff will not be able to learn effectively. Training initiatives must be customized to meet learning capacity based on the time constraints of those being trained. Training that can be readily managed by participants will have a much higher likelihood of success than those that layer additional burdens on already overworked staff.
- Clinical staff committed to quality improvement
    - Training should instill the concept that quality improvement, whether it be reviewing processes, updating skills, or ensuring that skills are maintained, is a crucial part of delivering high-quality patient care.
- Participants committed to spreading education
    - LMIC staff that acquire new skills through training programs should be encouraged as much as possible to disseminate their knowledge to colleagues. This can be within their own clinics as well as other clinics. This educational model results in fast, widespread dissemination of training.
    - It is important to spend adequate time choosing a department that will most likely fulfill this requirement for sustainability
    - Verbal and written commitments from clinical staff, department leads, hospital administration and, in some cases, the LMIC Ministry of Health are needed to confirm the willingness of the department to use the training given to them to then train others from the region (for low, or ideally no, cost)
    - A list of example questions that may be useful to ask potential LMIC participating departments are as follows (note: this list is not exhaustive but provides examples of issues that will need to be considered. In addition, the types and number of questions could change depending on the location and scenario):

- What is the infrastructure like? Do you have access to reliable internet and electricity? Do you have information technology and engineering support?
- What clinical equipment do you have? What other equipment/software do you have?
- What treatments do you perform? (This is to assess the level of complexity and give an indication of the local patient cohort.)
- How many patients do you treat per day? How many patients does your city, country, or region need to treat daily?
- Are your staff interested in entering an educational training program? Are any staff members opposed to it?
- Is your department/hospital administration supportive of staff participation in an educational training program?
- Is the Ministry of Health supportive of an educational training program?
- What do you want to learn?
- What is your staffing model? Is this enough staff to maintain the clinical duties as well as learn new methods of clinical practice? How much time can you dedicate towards an education and training program daily, weekly, monthly? For how long can you participate in an education and training program?
- Is your staffing model sufficient to support an increase in throughput if there are treatment efficiencies created?
- Is your staffing model sufficient to support an increase in complexity of certain treatments?
- Are you willing to train different staff than you may typically use to do certain jobs?
- Is your staffing model sufficient to introduce regional training at your facility?

9.4.2 Nonhuman

Nonhuman resources must also be considered when educational initiatives are being planned. Educational programs should either ensure that adequate nonhuman resources exist in the clinics where training is going to take place or are designed to accommodate the lack of those resources.

9.4.2.1 Technology

Technology has always offered new and interesting ways to develop and deploy educational initiatives. The following are examples of how current technologies can be used to enhance educational programs.

- Video—video-based training has many advantages. Trainers can demonstrate and discuss specific skills in a format that allows trainees to watch the training as many times as needed for them to become proficient. The most impactful part of video-based learning is that it is persistent in a way that in-person training is not. The primary downside of video-based learning is that the trainee cannot interact with the trainer and so cannot ask specific questions of the trainer in real time [10].

- Images/diagrams—images, and especially diagram-based instructions, can be incredibly useful in helping trainees develop skills and knowledge. The primary benefit of images is that they can be made language neutral, meaning that one set of images can be used in many different settings. As videos, they are also persistent in a way that in-person training is not. The primary downside to image-based learning is that the trainee cannot interact with the trainer in real time.
- Video conference—video conferencing offers the flexibility to meet with trainees on a regular basis without the expense of travel and lodging. They also enable screensharing, so skills that are primarily developed on computers are particularly well suited to video conferencing [11].
- Automation/machine learning/artificial intelligence—there are now many commercially available clinical automation tools. These can be effective in improving efficiency and standardization treatments, but trainers must ensure that trainees have sufficient background knowledge and skills to notice when automation produces an undesirable outcome and to be able to manually remedy it.
- Web-based platforms to house training resources and host remote training sessions are critical. These may require technological input from external providers.
- Web-based clinical practicum—a trainee in an LMIC may practice any clinical scenario that can be simulated on a PC and receive real-time critique, typically in the form of a quantitative comparison to an expert's work on the same problem. This is an extremely efficient mechanism for creating safe clinical professionals as they can practice real clinical skills (something that many clinicians in LMICs desire more than anything else).

9.4.2.2 Finances

Financial considerations must be taken into account when assessing the goals and priorities of LMIC education. There must be 'buy-in' and effective marketing to ensure health and government authorities are supportive of the proposed training, and any program must be able to demonstrate effectiveness in order to sustain financial support.

In any educational or training endeavor, finances are important to consider, even for philanthropic endeavors. Sustainability of any training program depends on financial feasibility.

There are three primary methods of generating income for volunteer-based enterprises: fundraising, grants, and partnerships [12].

- Fundraising can often be 'hit and miss' depending on the ability of an organization to promote its purpose. When fundraising, it is important to research the pros and cons of various methods of fundraising, and then use or fine-tune a specific concept that may work best to maximize income for the organization, organizational objectives, and goals.

- Grants can generate a significant amount of money if they are pitched correctly. Educational grants are typically harder to come by than research grants and thus it is important to keep in mind that funding bodies will typically want to support a program that can show a measureable increase in patient life expectancy or quality of life. Ideally educational grant applications should be able to reduce the cost per patient life to as close to zero as possible. However, treatment equipment and personnel are expensive and patient outcome data may be scarce in LMICs. Thus, in creating an effective argument for cost per patient, it is important to highlight how the program will positively affect three areas: patient throughput, sustainability, and reduction on the overall national healthcare burden. Improving patient throughput can be easily demonstrated, as a central goal of any training program should be an enhancement in treatment efficiency (e.g. being able to treat more patients per staff/equipment). Sustainability can be demonstrated by providing a creative approach to training program. By using the training program to create regional training centers, then the number of participants provided with training multiplies over the years, thus dramatically reducing the cost per patient of the program as the years go on. And finally, the reduction in national healthcare burden due to improved quality of treatment is a major factor in reducing the cost to patient and may provide extended benefits to other medical specialties not associated with the particular training program. This value is harder to quantify, but efforts should be made to highlight the domino effect on other medical/social services.
- Partnerships can be a long-term financial solution for global health education programs. The primary motivation behind any successful partnership is creating mutual benefit. An entity will be much more willing to donate money if they will benefit from the endeavor as well. However care should be taken to ensure partnership does not become a business partnership with the donating entity being the sole beneficiary of the success of the educational and training program. This can pose legal concerns for volunteer-based organizations. Typically, when a partnership proposal is thoughtfully written, emphasizing the benefit to both those in the partnership and participants in the program, the funding organization is more likely to enter discussions with the intention of coming to an agreement that works for everyone.

9.5 Practical implementation of a global health training program

A successful global health training program incorporates more than just education for local clinical tasks and quality improvement measures. It also provides education for leadership of future training in departments within the region. Often different teaching methods must be used to distinguish the two training pathways.

To create trainers, one must not only teach how to perform a clinical task, but also teach how to train someone else to perform that clinical task. Standardized methods of training can be implemented at an LMIC regional training facility, and new trainers should be taught how to use those methods. A highly scripted program will ensure the new trainer does not need to focus on how to teach, freeing up more time to use toward active mentorship. To further lessen the teaching burden on LMIC clinical staff, methods should be employed to reduce the in-person component of training as much as possible, noting that most of the knowledge can be imparted via text and video lectures, presentations, and demonstrations. A local trainer is necessary to fill any learning gaps, clarify more complex issues, and observe/assess clinical skills of those they are teaching.

The following are examples of methods for structuring a regional training center to relieve much of the burden of teaching from the (typically) busy clinical staff (note: a full analysis of the local setting and staffing model must be done to design the most appropriate training program for each facility):

1. Create an online platform to train students on a didactic portion of the curriculum. This platform should be carefully planned to suit the unique needs of the local audience. A combination of video, written, PowerPoint, Prezi, etc. lectures should be prepared to suit the varying learning needs of each participant. The length of each lecture or presentation should be sufficient to impart necessary knowledge but not so long that the participant loses focus or skips through portions of the lecture. If the information is both easy to understand and taught in a very clinical way, so that the principles can go from lecture to clinical practice with very little effort, then the participant will be much more likely to watch and pay close attention to the presentation. The lectures should be clear with no ambiguity that would necessitate additional research and learning by the participant. Different methods could be employed for delivery of the lectures, ranging from YouTube to a Moodle website. However, emphasis should be placed on creating a structured platform where sequential ordering of the lectures and proof of completion is inherent. In addition, there should be some form of assessment either during or after each lecture to ensure comprehension and help safeguard compliance. After completing a test, detailed answers to the questions should be made available to the participant, which can then be used as a summary of key clinical concepts. Participants are more likely to review the questions and answers and key clinical concepts summary rather than watching an entire lecture again. This online, didactic portion of the program could be initiated prior to any onsite visits by the participant or trainer. The regional trainer should still be available to answer any questions from participants if clarification is needed throughout the didactic portion of the curriculum.
2. An online platform could also be employed as a preliminary practicum component of the curriculum. Clinical staff commonly request to see demonstrations of how to do things in real clinical practice. Ideally, this will be performed by the participant under supervision of the trainer in a real clinical scenario on real equipment or real patients. However, there is value in tutorial videos of clinical tasks that participants can view multiple times

prior to undertaking the task themselves (under supervision). Some videos may be long demonstrations and other clinical scenarios might be best suited to have multiple, or even dozens of short (15 s to 5 min) videos to demonstrate the decision making process on a myriad of clinical scenarios, including what, why and how something is done.

3. Clinical documents incorporating protocols/procedures/guidelines/spreadsheets should be integrated within the training program. This has proven to be a successful and efficient way to ensure clinical implementation of concepts and techniques taught and will ensure standardization of processes throughout the LMIC as these documents can be readily shared.
4. An on-site, clinical practicum should be incorporated into the curriculum. This should include methods for demonstrating and assessing completion and comprehension for clinical tasks. Standardized methods of quantitative and qualitative assessment should be created and taught to the trainer. The practicum should be easily followed by the participant to minimize the oversight of the trainer when scheduling tasks to be completed.

It is important to regularly follow-up with the regional training facility to ensure that training is running smoothly, and to help local staff overcome any unexpected obstacles encountered during implementation and running of the training program. Access to any and all teaching and education platforms should be maintained for the regional training facility and continued recognition of the trainers' efforts can also help lead to improved inertia of the program and overall sustainability. Figure 9.1 gives an example of a possible timeline of a collaborative training program covering multiple LMICs at once with a large team of educators. This example is based on the Rayos Contra Cancer training model. Within this model items may be compressed or expanded as needed. The timeline is designed to be longer than needed in order to make contributions from volunteers more manageable and sustainable. This timeline is also specifically designed for the first iteration of a training program, and hence some items can, and will, be deleted or modified for subsequent iterations. For example, documents are prepared on the first iteration of the program, so this section can be deleted for all succeeding iterations of the program in different countries or regions (with the exception of language translation at times).

9.6 Conclusion

Deficiencies in equipment and staffing in many LMICs is well-known and thus clinical efficiency needs to be taught in addition to clinical efficacy and safety. As with most scientific endeavors, it is important to first perform a literature review to learn what others have done in terms of clinical efficiency on the treatment modality being taught. Do the concepts exactly transfer to the unique local setting of the LMIC? If not, how can those principles be adapted to work more effectively in a training program?

Some concepts may translate quite easily from developed countries to LMICs. However, others don't. Standardizing treatment procedures for all patients removes the need to drastically change procedures and expedites treatment time (due to the

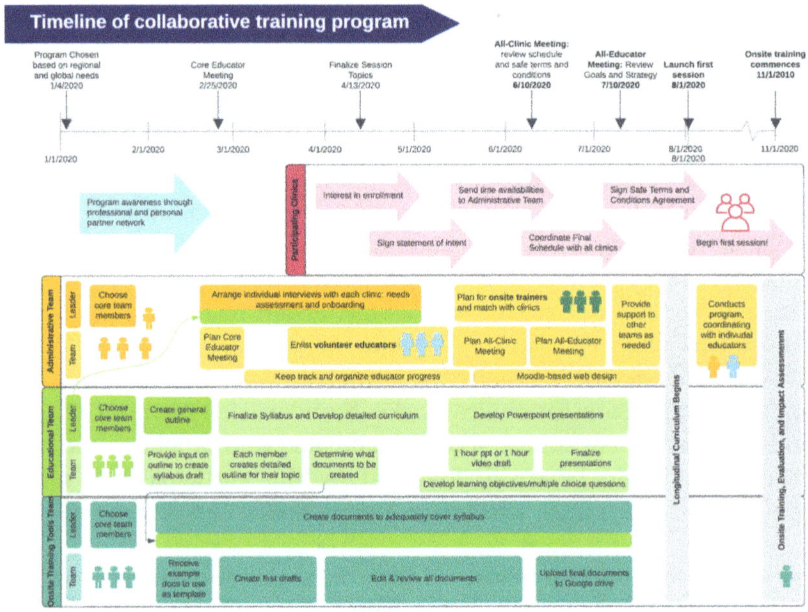

Figure 9.1. Example timeline of a collaborative training program (Rayos Contra Cancer, 2020).

repetitive nature of the treatments). In many LMICs, there is a great need to treat as many patients as possible, which means that individual, customized treatments that may be given in developed countries are not in the best interests of the LMIC clinic or patient population. Higher level, more complex treatments can still be taught and administered in LMICs, but careful patient selection to determine which patients that will receive the greatest benefit is a possible compromise in these instances. If this occurs, then LMIC departments will need to be proficient in both standardized and specialized treatment techniques, increasing the training burden on both the participants and educators.

The education of health professionals in LMICs must have a clear objective: to ensure availability of a workforce that is adequate in number and has the competencies and skills that correspond to the needs of the population it serves. To ensure the objective is being achieved, any education program must be able to assess participants and be assessed itself with respect to the quality, quantity and relevance of program graduates in LMIC clinical departments. This is the final challenge in development of educational packages and is often the hardest component to undertake in an effective manner. Statistics, qualitative and quantitative analysis plus programmatic assessment will be helpful in validating a program is meeting the training objective and will provide valuable evidence for quality improvement in future programs.

As has been identified in this chapter, many paradigms exist to educate and train medical staff in LMICs. Each method has pros and cons, and sometimes different methods can harmonize to create an even better training program. However, in the end, the most critical component of any training program is thoughtfulness.

The educator, the training program director, the training program administrator, or anyone else involved in any training endeavor must spend time to understand the needs of the trainee, as well as how that trainee will best retain the knowledge being taught. Care must be taken to assess what the trainee needs to know, what the trainee wants to know, and what the trainee doesn't even know that he needs to know. Care must be taken to assess how the trainee will best learn, including how knowledge flows in the clinic for that individual, and hence how knowledge flow should be replicated in a training program on those same clinical concepts. Thoughtfulness is key. One cannot merely create a lecture, but create a functional mechanism for knowledge to be retained in clinical practice.

References

[1] Asirwa F C, Greist A, Busakhala N, Rosen B and Loehrer S P J 2016 Medical education and training: building in-country capacity at all levels *J. Clin. Oncol.* **34** 36

[2] Frehywot S, Vovides Y, Talib Z, Mikhail N, Ross H, Wohltjen H and Scott J 2013 E-learning in medical education in resource constrained low-and middle-income countries *Human Resour. Health* **11** 1–15

[3] Barteit S, Guzek D, Jahn A, Bärnighausen T, Jorge M M and Neuhann F 2020 Evaluation of e-learning for medical education in low-and middle-income countries: a systematic review *Comput. Educ.* **145** 103726

[4] Joos U, Klümper C and Wegmann U 2022 Blended learning in postgraduate oral medical and surgical training-an overall concept and way forward for teaching in LMICs *J. Oral Biol. Craniofac. Res.* **12** 13–21

[5] Tansley G, Bailey J G, Gu Y, Murray M, Livingston P, Georges N and Hoogerboord M 2016 Efficacy of surgical simulation training in a low-income country *World J. Surg.* **40** 2643–9

[6] Carrasco H, Fuentes P, Eguiluz I, Lucio-Ramírez C, Cárdenas S, Leyva Barrera I M and Pérez-Jiménez M 2020 Evaluation of a multidisciplinary global health online course in Mexico *Global Health Res. Policy* **5** 1–11

[7] Moreira D C, Gajjar A, Patay Z, Boop F A, Chiang J, Merchant T E and Qaddoumi I 2021 Creation of a successful multidisciplinary course in pediatric neuro-oncology with a systematic approach to curriculum development *Cancer* **127** 1126–33

[8] Reisman J, Arlington L, Jensen L, Louis H, Suarez-Rebling D and Nelson B D 2016 Newborn resuscitation training in resource-limited settings: a systematic literature review *Pediatrics* **138** e20154490

[9] Deci E L and Ryan R M 1985 *Motivation and Self-Determination in Human Behavior* (New York: Plenum)

[10] Pawlicki T, Brown D, Dunscombe P and Mutic S 2014 i.treatsafely.org: an open access tool for peer-to-peer training and education in radiotherapy *Med. Phys.* **2** 407–9

[11] Beck A, Nadkarni A, Calam R, Naeem F and Husain N 2016 Increasing access to cognitive behaviour therapy in low and middle income countries: a strategic framework *Asian J. Psych.* **22** 190–5

[12] Ayaz O and Ismail F W 2022 Healthcare simulation: a key to the future of medical education —a review *Adv. Med. Educ. Pract.* **13** 301–8

www.ingramcontent.com/pod-product-compliance
Ingram Content Group UK Ltd.
Pitfield, Milton Keynes, MK11 3LW, UK
UKHW051849100926
8556IPUK00005B/534

9 780750 337496